ATTENTION:
Due to extreme economic conditions,
there will be no laughing in the workplace.
Thank you.
– THE ECONOMY

DEFY THE ECONOMY

You no longer need to be at the mercy of the economy.

We can help you become an Adaptive Enterprise.

CAP GEMINI
ERNST & YOUNG

Management Consulting   Systems Transformation   Information Technologies   www.DefyTheEconomy.com

Innovation

Harnessing Creativity for Business Growth

Consultant Editor:
Adam Jolly

KOGAN
PAGE

Publisher's note

Every possible effort has been made to ensure that the information contained in this book is accurate at the time of going to press, and the publishers and authors cannot accept responsibility for any errors or omissions, however caused. No responsibility for loss or damage occasioned to any person acting, or refraining from action, as a result of the material in this publication can be accepted by the editor, the publisher or any of the authors.

First published in Great Britain and the United States in 2003 by Kogan Page Limited

120 Pentonville Road
London N1 9JN
www.kogan-page.co.uk

22883 Quicksilver Drive
Sterling VA 20166–2012
USA

© Kogan Page and Contributors 2003

British Library Cataloguing-in-Publication Data

A CIP record for this book is available from the British Library

ISBN 0 7494 3627 1

Typeset by Saxon Graphics Ltd, Derby
Printed and bound by Scotprint, UK

Contents

Preface
Design Council   *ix*

Foreword
Andrew Warren, Vice President and Head of Sales &   *xi*
Marketing, Cap Gemini Ernst & Young

Section 1: Inspire

1.1 Strategic commitment – the foundation stone   **3**
for innovation
Dan Rixon, Executive Consultant, Cap Gemini Ernst & Young

Case studies:
Renishaw
JCB
BAE SYSTEMS

1.2 A culture of innovation   **13**
Margaret de Lattre, Managing Consultant,
Cap Gemini Ernst & Young

Case studies:
AA
Dyson
ITS Technology

1.3 Stretching performance   **23**
Michael Riding, Managing Director, Lloyds TSB Corporate

Case studies:
Creature Labs
Dundrum Bay Oyster Fishery

1.4 Where people come first   **31**
Professor Alec Reed CBE, Founder and Chairman,
Reed Executive plc

Case studies:
Dollond & Aitchison
TR Europe

1.5 **Learning and development** 39
Vince Darley, CEO, EuroBios UK Ltd

Case studies:
Creature Labs
Thames Water
Select Software

Section 2: Create

2.1 **Teams and networks** 51
Garrick Jones, Executive Consultant,
Cap Gemini Ernst & Young

Case studies:
TR Europe
Linear Drives Limited
ITS Technology

2.2 **Leadership** 61
Alisdair Mann, Vice President, Cap Gemini Ernst & Young

Case studies:
A&R Cambridge
JCB
AA

2.3 **New ways of working** 71
Nigel Crouch, Senior Industrialist, Innovation Group, DTI

Case studies:
Dollond & Aitchison
TWI

2.4 **Risk management** 79
Adrian Blumenthal, Development Director, Rethinking
Construction

Case studies:
JCB
BioProgress Technology

2.5 **Intellectual property** 87
Anthony Murphy, Director of Copyright, The Patent Office

Case studies:
Gorix
Renishaw

2.6 Funding innovation 97

Michael Riding, Managing Director, Lloyds TSB Corporate

Case studies:
BioProgress Technology
Marks & Spencer
Peabody Trust

Section 3: Connect

3.1 Partnering with customers 109

William Mellis, Vice President, Cap Gemini Ernst & Young

Case studies:
BAE SYSTEMS
Creature Labs
A&R Cambridge

3.2 Information systems: innovation – it's changed! 119

Andy Mulholland, Chief Technology Officer,
Cap Gemini Ernst & Young

Case studies:
Waitrose
Bank of Scotland

3.3 Supply chains: collaborate to innovate 127

David Atkinson, Senior Consultant, Simmons Dickinson

Case studies:
Sainsbury
Marks & Spencer

3.4. Design and innovation 135

Jane Pritchard, Innovation Services Leader, IDEO

Case studies:
Dyson
Linear Drives Limited

3.5 Working with rules and standards:
rewriting the rules on regulation 145

Nigel Crouch, Senior Industrialist, Innovation Group, DTI

Case studies:
Gecko
Microsense Systems Ltd

Appendix: living innovation – self-assessment 153

Does your bank have the expertise to focus on your unique needs?

Every business is unique – with its own challenges and opportunities.

With Lloyds TSB Corporate, you'll have your own dedicated relationship manager who can be involved with your business as much or as little as you choose – providing you with the exact level of support and expertise you need.

It might be that you need someone with in-depth knowledge of your market sector – including current trends and the issues you might be facing as a result. You might be an organisation with complex financial needs that requires advice in areas such as Acquisition Finance or Capital Markets. Or you may simply require a manager who can deliver straightforward financial support, through a fast, efficient banking service. Whatever your requirements, you can be sure the expertise is on hand to deliver answers tailored to your needs.

To find out more about the benefits of Lloyds TSB Corporate or to arrange a meeting, call Andy Jeffries on 0117 923 3872 or email corporate@lloydstsb.co.uk

 **Corporate**

Preface

Few UK businesses can develop and grow sustainably by making and selling commodity products. Some rail against this – but it is a reality. And excitingly it's a reality that often leads to greater success for businesses that embrace it by offering added value products and services.

Design is a key factor in adding that value and differentiating businesses from local, national and international competition.

Design Council research (National Survey, 2002) demonstrates a clear link between the way a company understands and uses design and its business performance. For example, 91 per cent of rapidly growing companies use design, compared with 49 per cent of businesses overall.

Introducing new products or services is crucial to business growth. So it is not surprising that 83 per cent of growing UK businesses have done just that in the past three years. Perhaps the surprise is that some companies are not even trying. The evidence is that only 17 per cent of companies that had stayed the same size had introduced anything new. You cannot help wondering how long they will be in business.

This book breaks down innovation into three parts: Inspire, Create, Connect. This reflects the Design Council and DTI Living Innovation project, which analysed the processes used by a number of the Design Council's Millennium Product companies. You can find more about this and an online self-assessment at www.livinginnovation.org

At its crudest, innovation is simply about doing something new. What design brings to the party is an ability to connect bright ideas to real customer needs: producing great products and services that delight customers.

Growing companies recognise a clear connection between design and sales: 75 per cent of them say that design played a part in their sales growth. Sadly, most companies that hadn't grown also could not see a role for design in increasing their sales.

Harry Rich
Director – Business
Design Council
www.designcouncil.org.uk

contributor profile

andrew warren

vice president

Andrew is Vice President and head of sales and marketing at Cap Gemini Ernst & Young.

The Cap Gemini Ernst & Young Group is one of the largest management and IT consulting organisations in the world. The company offers management and IT consulting services, systems integration, and technology development, design and outsourcing capabilities on a global scale to help businesses continue to implement growth strategies and leverage technology. The organisation employs around 55,000 people worldwide and reported 2001 global revenues of more than 8.4 billion euros.

Foreword

Business is obsessed with innovation. Companies operating in expanding markets are always looking for something that will enable them to continue to grow. Companies operating in volatile or difficult markets believe they have to find new ideas in order to maintain or improve their market share.

Most industries want to believe that creativity can deliver a magic answer, a silver bullet. They pray to find the next big thing in the expectation that they can deliver it to the market a little bit quicker than anyone else.

We at Cap Gemini Ernst & Young are also obsessed with innovation. We don't, however, believe in magic answers, silver bullets or prayers. We may spend a lot of time talking about the ergonomics and psychology of creativity and innovation, but we do it for a reason. We do it because we want to turn ideas into facts.

We have invested heavily in developing new, radically different processes, methodologies and workplaces. As a result, we consider ourselves to be world class at industrialising creativity.

innovateUK, our innovation and solution acceleration centre on Wardour Street, London is about creating barrier-free environments within which people can be more creative and produce more innovative solutions to the problems they face. This increases the certainty of those solutions being implemented.

Having implemented an idea, the next step is to sustain it. Most people will take one creative idea or innovative solution and they will flog it to death. Truly gifted people, on the other hand, are already thinking about the next step before the idea or solution is even halfway through its lifecycle.

Good businesses obsess about innovation constantly. Great businesses obsess about it constantly and do something about it.

Andrew Warren

1 Inspire

1.1

strategic commitment – the foundation stone for innovation

'without strategic focus and commitment creativity may surface naturally but innovation will not be delivered'

Dan Rixon, Executive Consultant, Cap Gemini Ernst & Young

contributor profile

dan rixon

executive consultant

Dan Rixon is an Executive Consultant at Cap Gemini Ernst & Young. He works with leaders and organisations to deliver innovative solutions across an extensive breadth of industries.

His unique combination of competencies include leadership coaching, transformation management and collaborative work design. His ability to deliver innovation results in his clients achieving higher levels of commitment to bolder more innovative solutions than organisations typically achieve.

Dan has developed these capabilities from over 17 years of board level strategy and transformation consulting and has lead the Accelerated Solutions Environment for the last 5 years. He is recognised as Europe's most experienced facilitator of collaborative and innovative work programmes and events. Prior to this role he held a leadership position in Cap Gemini Ernst & Young's Organisational Change Practice and has managed global transformation programmes.

strategic commitment – the foundation stone for innovation

Almost all would accept that innovation cannot be fostered in an organisation without strategic commitment. But what should a leader, organisation or team commit to? Is it to a vision, a culture, a specific result or innovation itself? And once you know what to commit to, how is commitment demonstrated and how is it improved?

First we must consider what innovation is and actually more importantly what creativity and the creative process is.

What is the difference between innovation and creativity?

Creativity is a process of developing and expressing novel ideas that are likely to be useful.

The end result of the creative process is an innovation, where innovation is the embodiment, combination, and or synthesis of knowledge in novel, relevant, valued new products, processes or services.

When Sparks Fly – Dorothy Leonard, Walter Swap

Commit to a creative process designed to deliver innovation

This definition suggests that if an organisation wants innovation they must commit to a creative process. Renishaw's creativity is institutionalised in their product design process and for both Renishaw and JCB the creative process is embedded in the culture with significant importance placed on idea generation and collaboration in problem solving. More controversially it suggests that having the word innovation as a core value like BAE SYSTEMS is not enough and that the creative process established around the small and flexible project team was more relevant in producing the innovation.

Create and commit to a strategic focus for creativity and innovation to succeed

It is vital that leaders give their organisations and teams the strategic focus for the desired creativity and innovation. Is the innovation needed to re-create a market or product? What degree of innovation is needed to succeed and what level of risk is acceptable? Is it acceptable to redefine the business model? Without an insight into these boundaries ideas are likely to be too readily considered too difficult. This direction need not be detailed. It should merely create and bound a focus for innovation.

McMurtry creates the focus for the Renishaw organisation very simply by directing the organisation to 'Protect your products…and back this up by defending them'. An employee is now empowered to be as creative as they can in defending the business from its competitors. More importantly the focus now allows the internal implications of the innovation to be managed more easily creating an environment where the natural response to an idea is 'can it be made to work?' rather than 'it won't work'. Without strategic focus and commitment creativity may surface but innovation will not be achieved.

The strategic focus driving success in the JCB and BAE SYSTEMS examples is product and project specific, targeted at specific market demands. Both teams were so inspired by the focus they were given and the opportunity they had to change their markets that they not only reinvented their products they reinvented the way they worked.

Create an environment where individual and organisation confidence is increased by participating in the creative process

Having established a creative process and focussed it into the desired space for innovation it now needs to be led to success. Leaders must create an environment where individuals are confident that their radical or challenging idea has the potential to be valuable to the business. Leaders need to commit to supporting individuals who contribute to the creative process.

Imagine now walking down a busy street in EuroDisney and meeting Mickey Mouse. Your engagement with Mickey is fun, Mickey is being remarkably creative acting the role of a mouse and the encounter is full of enthusiasm and imagination. Mickey moves on to the next family with even more enthusiasm and creativity.

Consider now the same interaction, this time in your local high street and Mickey isn't wearing a costume. You reaction is now likely to be one of surprise and probably embarrassment. Mickey is initially full of enthusiasm but after another two or three similar reactions from others in the high street he gives up and loses confidence.

It is an extreme example but a simple one that indicates the importance of creating the right environment and conditions for creativity to surface and then develop. Creativity in an unmanaged environment creates the situation where one person's creativity is another person's stupidity, driving a rapid lack of confidence in the creative process.

Sir Anthony Bamford's support for the JCB Teletruk project, 'If the strategy is right we will find a way to get the numbers right' increases the confidence of his team to achieve an innovation and overcome organisational issues. McMurtry increases the confidence of his workforce by discussing innovation at employees' desks. In addition to the merits of his inspirational communication, by visiting employees and project teams in their own environment he is also legitimising their workspaces as creative spaces embedding creativity further into Renishaw's culture.

In summary
Strategic commitment is vital for innovation to be achieved. Without strategic focus and commitment creativity may surface naturally but innovation will not be delivered.

Renishaw is almost aggressive in its approach to innovation.
Renishaw turns its back on established market sectors where competition drives prices down. Instead it spends 12 per cent of its profits on research and development, focusing on niche markets in inspection technology where rivals are few and margins are high. Its mission is to 'design, manufacture and supply metrology systems of the highest quality and reliability to enable customers worldwide to carry out dimensional measurements to traceable standards'.

case 1

close inspection

Its analogue probe (SP600M), for instance, places the company at the forefront of high-performance, high-speed measurement devices, enabling surface data to be collected from 3D objects with greater accuracy and lower contact force than ever before. It was one of the eight Millennium Products awards that Renishaw received from the Design Council, more than any other company.

David McMurtry, Renishaw's founder and its chairman, is very much the company's driving force. An innovative culture is nurtured by his dedication to growth, R&D, niche markets and commitment to protecting intellectual property. 'Protect your products,' says McMurtry, 'and back this up by defending them.'

Renishaw is almost aggressive in its approach to innovation. A high priority is attached to creating a culture of close-working, multi-skilled teams and strong inter-action between staff and management. Employees are actively encouraged to communicate, provide feedback and develop new ideas.

Renishaw has an advantage in creating a close-knit sense of community: 800 of the company's 1100-strong workforce are based in Wotton-under-Edge, which is also McMurtry's home town. As a major employer in the area, Renishaw has fostered strong links with the local school and the wider community. This commitment to community is reflected in levels of staff satisfaction; annual staff turnover stands at just two to three percent. The emphasis on team working is a major consideration – from the recruitment process through to the corporate structure itself.

McMurtry – and his philosophy of commitment and empowerment – is the driving force behind the company's culture. He claims to have appointed clones as heads of each of his directorates to instil his personal values in the workforce!

Divisions are responsible for managing their own resources, employees are encouraged to feed into projects and suggest new ideas and, vitally, they are not penalised for making mistakes. McMurtry spends much of his time at employee's desks, discussing their perceptions of designs and projects.

case 2

numbers follow strategy

Strategy and reality appeared to be at odds with each other.
The chairman of JCB, Sir Anthony Bamford, was seeking to expand from construction and agricultural vehicles into the industrial handling sector. Analysis showed this was dependent on manufacturing a forklift truck to build volume and establish a presence in the market sector.

Such a move would not be easy. The market for forklift trucks is both mature and fiercely competitive. Forty different companies in Europe are already manufac-turing forklift trucks. There was no way a 'me-too' product could be profitable. Product innovation was essential.

JCB quickly rose to the challenge. The company has a culture in which creative new ideas can flourish. Around a thousand staff suggestions every year evolve into a hundred product ideas. These are distilled into 15–20 project proposals, with around half reaching production.

This reflects the positive attitude to innovation taken by the company's chairman, Sir Anthony Bamford. He provides top-level support for creativity and sees sensible risk-taking in a positive light. Though the initial projections suggested that the proposal for a new forklift, the Teletruk, would not be in profit for several years, he took the view that, 'if the strategy is right we will find a way to get the numbers right'.

Within four weeks a dedicated 12-strong team had produced a business plan. By moving personnel into previously unfamiliar areas of work, JCB ensured team members were not blinkered by prior experience and took a completely fresh approach to the initiative. They adopted a radical approach to forklift truck design involving use of a single telescopic arm from existing JCB digger machines. This has far stronger user appeal than conventional forklift trucks.

Confidentiality agreements signed by local companies, partners and suppliers allowed a great deal of expert input into the design of the new 'Teletruk'. Continuous attention to the life costs of the project allowed the team to 'get the numbers right' and improve the gross margins from those originally projected.

JCB Industrial met its launch date target and its Teletruk went on to win various industry awards and be selected by the Design Council as a Millennium Product. Early sales are not only on track to meet targets; within the first month of production, JCB's chairman had already sent the team back to work on planning how volumes could be increased to allow expansion into new markets.

Innovation has become one of the company's five key values – perhaps it is the most important.
BAE SYSTEMS almost missed out on the booming demand for gyroscopes. But it re-organised itself and set up a unit dedicated to developing a market-leading silicon gyroscope. In the process, it not only won Millennium Product status for the product, but also changed its culture forever and established innovation as one of the company's five core values.

case 3

BAE SYSTEMS

It first considered replacing conventional rotating gyroscopes with solid state ones in 1992, when it was searching for a more rugged device to withstand the shock of launch in mortar shells. The military requirement came to nothing, but commercial demand for smaller, less expensive and more robust gyroscopes was increasing. Unless BAE SYSTEMS responded, its share of the market could disappear.

BAE SYSTEMS developed a specification for an easy-to-use, small, rugged gyroscope with low power consumption, which would work at least as well as anything else on the market. No products existed that matched these criteria, so the company was really acting on a gut instinct, according to Colin Fancourt, head of solid-state sensors. It knew that this was the way to go, but production costs would have to fall a hundredfold. That is where the silicon idea came into being.

The managing director of BAE SYSTEMS Plymouth Business Unit, Nigel Randall, who is an ex-engineer, had faith in the project and was always a keen supporter. 'But we had to convince his board and our sponsors at HQ,' says Fancourt. 'So we spent modest sums on formal risk assessments, marketing analysis and engineering analysis. We had to pass through a few gates.'

BAE SYSTEMS also discovered that its competitors were working on a quartz version, using the technology established in wristwatches. Its view was that a cost-effective quartz system could not be developed that was tough enough to handle temperatures in both the Arctic and in Death Valley or which could withstand powerful shocks from stone impact and engine vibration.

In choosing silicon over quartz, BAE SYSTEMS opted for an area that was not heavily protected by intellectual property, enabling it to take a new direction. Its silicon technology has subsequently been thoroughly patented in all relevant markets.

There was still no guarantee that this was the right decision. Nor could BAE SYSTEMS be sure that it could engineer silicon to the dimensions required, as micro machining was in its infancy. The silicon route was much higher risk. 'With the benefit of hindsight, the choice is easy,' says Colin Fancourt, head of business for solid-state sensors. 'Silicon is the better technology. At the time, if you'd been a betting man, you'd have put your money on quartz because it was proven technology.'

Solid State Sensors was set up as a separate business area, creating a team of ten people with all the relevant skills – development engineering, manufacturing and research. Sixty people are now involved. 'We are an entrepreneurial part of BAE SYSTEMS. We are deliberately small. It helps us to stay flexible and maintain creativity,' says Colin Fancourt. This approach has been instrumental in creating a new culture at BAE SYSTEMS, which fosters and encourages innovation.

This policy has been extended through partnerships. BAE SYSTEMS created the design for the gyroscope, but consulted Nottingham University on the theory and called on Cranfield and Loughborough universities for manufacturing advice. It used Denbys International for marketing the idea to the automotive industry. It also formed a partnership with the Sowerby Research Centre in Bristol, a BAE SYSTEMS establishment, to conduct fundamental research on silicon.

BAE SYSTEMS was responsible for the product's design. Its manufacturing partner was Sumitomo Precision Products in Japan. Bridging the cultural gap between the two was the main difficulty in developing the project. Resolving these differences has given BAE SYSTEMS a benchmark for use with future design innovations. 'They accepted the need for speed. But we have also accepted the need for far more consensus,' says Colin Fancourt. 'We've learnt that from them. And we've also learnt to design for manufacture. They always design something with manufacture in mind. That lesson has been very important to us.'

The silicon gyroscope project has made BAE SYSTEMS at Plymouth less cautious as a whole, says Colin Fancourt. 'We are now far keener on innovation, more prepared to take risks. The success of this project has contributed to a wind of change that has blown through the company. We now actively encourage innovation. We are far less risk averse. The company chairman now makes awards for innovations each year. Innovation has become one of the company's five key values – perhaps it is the most important.'

1.2 a culture of innovation

'for any organisation pursuing a
vision and strategy with a focus
on innovation, creating and
sustaining a culture to support it
is a key foundation stone'

Margaret de Lattre,
Managing Consultant,
Cap Gemini Ernst & Young

contributor profile

margaret de lattre

managing consultant

Margaret de Lattre is a Managing Consultant at Cap Gemini Ernst & Young, responsible for big business client transformation programs in the UK.

She leads a team responsible for developing culture and behavioural change throughout their different levels of hierarchy.

With over six years as a consultant she has extensive experience in culture and behaviour in a wide range of organisations, both public and private sectors. Prior to this she was a change manager, with a particular interest in culture and behaviour.

a culture of innovation

Culture underpins the way an organisation works every day. It is the unconscious competence of a business. For any organisation pursuing a vision and strategy with a focus on innovation, creating and sustaining a culture to support it is a key foundation stone. For organisations trying to increase their focus on innovation, it is particularly important to recognise that without the reinforcement of changes in the underlying culture, they will quickly slip back into their old, familiar and less innovative ways of working.

Culture in organisations is driven by a complex series of interactions. Many managers when addressing the issue of culture treat it too simplistically. Publishing a statement that says the business values innovation, producing some posters to communicate it and describing some behaviours is all too often as far as many organisations go.

The following three case studies demonstrate how innovative companies intervene at many different levels to create and sustain a culture that supports their goals and objectives.

Vision
The old hobby-horse of the management gurus is still alive and well. Having something clear to aim for, behind which to focus the culture, is just as important here as with any other management topic. A huge topic like 'innovation' with its myriad of potential interpretations, can be too difficult a concept for many to grasp. The three

case study organisations that follow have helped people to understand their particular need for innovation by tightening their vision into something easier to grasp and more difficult to misinterpret. The AA focuses on customer service through flexibility, Dyson on 'can do', and ITS Technology on commercialising the best technologies. Though not explicitly stated in many instances, these themes are unconsciously embedded in their approach to everything they do.

Symbols

Communicating the focus on innovation throughout the business is an imperative. The AA helped its people to work in the new way by listening to their problems and then providing them with the information they needed to do things differently. Speeches, memos and posters will not be enough, however. The theme of innovation needs to be re-enforced in a broad range of symbols that communicate to all who come into contact with the organisation. The AA did not forget its suppliers and customers in its communication. Dyson's emphasis on informal dress (innovation is rarely formal) and its building design, not only communicate innovation visibly, but also enable innovation to take place through flexible working and freedom of movement, and associated ideas. The focus at ITS Technology on going beyond discussing R&D and turning communication to 'marshalling the value' is also significant.

Behaviours

Encouraging behaviours that foster innovation through skill building and role modelling is also evident, and making these innovative behaviours highly visible and laudable is important. ITS Technology has built close working and learning relationships with organisations at the leading edge of invention. James Dyson's own determination and perseverance, as well as his personal role and involvement in the participative process of ideas creation, is highly significant in fostering innovation at Dyson. The AA involves customers and employees in developing solutions. Another point of note is Dyson's focus on building the business from within, to create an internal momentum for innovation from learning new behaviours and skills within the team.

Reward and measurement

Valuing people for their contribution is always a good driver of performance, although financial reward is not always the key, as with Dyson. The areas the company tracks, measures and reports upon is critical, as this should be the driver for personal goals and objectives throughout the organisation. Dyson and ITS Technology report on R&D spend, a traditional corporate measure for innovation. They also track other familiar trends but do so with an eye on innovation, such as customer needs as a focus for R&D effort (ITS) and staff turnover as in indicator of internal opportunities from a creative atmosphere (Dyson).

Business context

Communicating, defining appropriate behaviour, role modelling, rewarding and measuring innovation is all to no avail if the systems, structures, policies and processes of the organisation are in the way. The AA fundamentally restructured its business to enable its people to work in a new way, from automating call centre activities and removing territorial boundaries, through to involving a customer/user community in the IT project group's design decisions. Flexibility to enable innovation in customer service is a recurring theme in their structural and procedural changes. Dyson and ITS Technology both cite the importance of recruitment policies and processes to enable the free flow of ideas. At the heart of Dyson's way of working is an informal and flexible face-to-face communication process, mirrored in a flat informal structure.

Each organisation is very different, and fosters innovation for different reasons. The AA, Dyson and ITS Technology case studies demonstrate how organisations of different types and sizes can help to create and sustain a culture of innovation by acting in the five areas that drive an organisation's ways of working.

Culture change was central to the national roleout of the plan.

The AA faced a growing threat to its roadside repair services. By addressing the issue from the customer perspective, it was able to transform the technological and operational structure of the service. The net effect has been to reduce average wait times to 33 minutes (which is less than that of the London ambulance service) and ensure 90 per cent of calls are attended within an hour. By reducing the number of call centres, the AA has cut costs, and by improving member service it has helped to ensure a growth in membership despite increased competition. The new vision of customer service has provoked interest from other emergency services and partner motoring associations in mainland Europe.

case 1

innovation on the roadside

The system introduced major changes for many employees and suppliers – particularly dispatchers, road patrols and those employees deciding which dispatchers to use. Dispatchers and road patrols were asked to be more flexible rather than work their own little 'kingdoms', and decisions that had previously been made by call centre controllers were automated. A culture change was needed to support these changes in business processes. Core to this culture change was increasing communication and understanding between management and users. Employees were supported and well informed. This culture change was central to the national rollout of the plan.

User, member, business and technological requirements were all taken into account in building the design. A customer/user community was set up alongside the IT project group. This community was crucial to any design decisions that were made. Work done with members in other parts of the AA identified key issues, which contributed to the overall design. Surveys, meetings, the use of external experts and an examination of technological products from third parties all underpinned this strategy.

The dispatchers and 600-strong patrol force were the hardest parts of the business to bring on board; their territorial boundaries were removed and a higher degree of flexibility was required. Before the workforce was convinced, the AA had to persuade patrol managers. This was accomplished by listening to the force's problems and supplying all the necessary information in the form of AA-produced books listing garages and services for the new areas, to overcome 'fear of the unknown'.

case 2

a restless company

'There is no such thing as a quantum leap…there is only dogged persistence and in the end you make it look like a quantum leap.'

The Dyson's DC03 vacuum cleaner is one of the most familiar products to be awarded Millennium Product status by the Design Council. Using two filters and no bag, the DC03 has a highly efficient filtration system. Its low wattage motor makes it ecologically sound as well as highly innovative.

The DC03 aimed to break into the European market, which did not buy upright cleaners, and the UK market, which had never bought slim-line, lightweight uprights. In one year, it became the UK's second highest selling cleaner and has been instrumental in growing the upright cleaner market in Europe.

17 per cent of the company's turnover is spent on Research & Development. 'Traditional domestic vacuum cleaners had not changed in 90 years,' says James Dyson. 'Dyson now holds 52[1] per cent of the market share – by value. While large firms fail to focus on R&D, the opportunity exists for young businesses to take advantage of their complacency.'

Led by James Dyson's visionary commitment to a 'can-do' ethic, the company employs 1,800 people in-house to do everything from advertising to patent applications. A flat, informal structure, the free-flowing communication of ideas and a close relationship with customers drives the innovation process on which the company is founded.

'There is no such thing as a quantum leap,' says James Dyson. 'There is only dogged persistence and in the end you make it look like a quantum leap.' Initially James Dyson was told that his products had no chance of success. He was turned down on the idea, the offer of a license and on finance. Displaying the determination and confidence in his people and products that now permeates his company, Dyson decided, 'No one else was going to do it. So, we had to do it ourselves.'

[1] GfK Lektrak May 2001 (by value).

Dyson has taken the ethic of 'doing everything ourselves' to heart; the business is built from within to an extraordinary degree. Wherever possible, people are employed in-house, creating a 'can-do' culture of self-confidence and empowerment. 'Solutions come from within,' explains Dyson. 'The heart of the company is creative and problem solving.' The company is fundamentally participative; free-flowing ideas are valued and the ethic of face-to-face communication is strongly endorsed.

Instead of emails and memos, employees are encouraged to discuss ideas in person. There are no formal presentations to James Dyson, simply the freedom to talk with him. The wearing of suits and ties is not encouraged. The flat structure is reflected in the building's design, which is modern and open, without boundaries between the departments.

Dyson employs no consultants; advertising, PR and technical development are all carried out in-house. Recruitment is therefore crucial; bright, young people with the 'right spirit' – and often without experience – are favoured. They will make mistakes, but they will also learn to recognise and rectify them. At induction, each employee builds a vacuum cleaner, empowering them with knowledge of all aspects of the business. Staff turnover is kept very low by an abundance of internal opportunities and a creative atmosphere. There are no bonus schemes; 'Anyone who needs a bonus to motivate them is not the right person to be working here,' says James Dyson.

'We learn how to be creative by doing it,' says Dyson. A helpline and data from in-house repairs are seen as invaluable media for learning and feedback is incorporated into the design of the next product. James Dyson values the concept of a 'restless company' in which the staff continually asks: 'What can we do to make it better?' The low staff turnover helps employees to accrue substantial experience; learning from experience is strongly encouraged. Good ideas developed in one programme are fed into the next.

High levels of customer loyalty and input enable the company to be customer-driven. The helpline gives customers direct access to Dyson employees. A large customer database has been created from the registration of the two-year guarantee. Weekly sales figures help to complete the picture. However, Dyson believes the company needs to be ahead of the customer; 'We are a restless company,' he says. 'As soon as a product is launched, we look to improve it.' Customers do not test prototypes until late in the development process. For new markets, forecasts are rough and are not taken as gospel.

'It's more a case of trying to stop some of the other directors talking about R&D than trying to identify new ideas.'

an ideas factory

ITS Technology was spun out of UMIST in 1997 to exploit a technique for producing real-time cross-sectional images. Instead of using X-rays, ITS uses electrical resistance to quantify the conduction and insulation of materials thousands of times a second. Such live analysis is beneficial, for instance, in identifying foreign bodies in processed food or in the North Sea where oilrigs are pulling up 90 per cent water and 10 per cent oil. Other applications are envisaged in pharmaceuticals, chemicals and minerals. The vision of the chief executive, Ken Primrose, is: 'a system in every factory'.

The company spends a lot of its resource on R&D. 'It has to,' says Primrose. 'That's what is going to make the company successful. One hopes the things we are investing in will bear fruit, the product is a very flexible instrument and we hope to be much more responsive to market needs.'

The company is based in Manchester with nine employees and a turnover of £0.5m. Primrose became involved after writing to UMIST looking to start a technology business. His objective was to bridge the divide between research and commerce. He provides the company's commercial drive, while the rest of the team is encouraged to improve on ITS's leading edge technology.

'It's more a case of trying to stop some of the other directors talking about R&D than trying to identify new ideas,' says Primrose. 'It's more a case of marshalling the value and finding which ones are best.'

All nine employees are highly qualified with MScs, MBAs, PhDs and professorships. They were taken on according to customer needs. The company is essentially composed of electronics engineers to make the product and develop the technology, and chemical engineers to solve problems of analysis.

There is a lot of published work on electrical resistance tomography, so a patent was out of the question. The intellectual property can only be protected by maintaining and developing know-how and by copyrighting software used in interpretation.

Personal development is mainly learning from other employees. A close relationship with UMIST is maintained and ITS sponsor research in other institutes and companies. ITS uses its local Business Link, particularly on the export side, and the company also finds the Business Excellence Model to be a helpful tool for change.

On launch, time to market was seen as critical but, two years on, there were no competitors using the technology. Other solutions are available and the company stays in touch with academics to ensure that it is aware of evolving technologies. Both technical and financial risks are seen as equally important and risk is managed in a financial model. Primrose's priorities are to: 'get sufficient funding to get core elements in place and then concentrate on sales and stay focused.'

**1.3** **stretching performance**

*'success will depend ultimately
upon communication of the
innovator's vision to those staff
whose support is needed to
realise his or her dream'*

**Michael Riding,
Managing Director,
Lloyds TSB Corporate**

contributor profile

michael riding

managing director

Michael Riding is managing director of Lloyds TSB Corporate, a business that provides a broad range of banking, advisory and financial services to the corporate marketplace. Michael joined Lloyds Bank in 1983 from Chemical Bank, New York. Senior roles within the retail side of the bank preceded his appointment as Head of UK Commercial banking in 1991. He assumed his current responsibilities in January 2000.

Part of the Lloyds TSB Group, the Corporate division delivers relationship banking expertise to companies with turnovers in excess of £2 million. In addition, and in support of corporate businesses, corporate banking is able to deliver a number of specialist financial solutions. These range from straightforward cashflow support via invoice discounting to businesses undergoing change and requiring substantial financing. Specialists from the Acquisition Finance, Development Capital, Capital Markets, Structured and Commercial Finance teams are able to understand these requirements and situations.

Lloyds TSB Corporate is based in London and Bristol with dedicated offices throughout England, Scotland and Wales, as well as a substantial presence in New York. On an international front, many of Lloyds TSB Corporate's links are worldwide, and there are extensive links with foreign banks, businesses and governments. All this enables Corporate to support business activities on a global basis. Its aim is to ensure that it matches its expertise to the requirements of the market in order for all its customers to meet their objectives.

stretching performance

Stretching performance can be viewed as a move by a business to develop its existing resources and competencies to provide itself with a competitive advantage. The mention of existing resources is noteworthy because recruiting staff and forging third party alliances do not necessarily have to form part of the stretching equation.

The examples of Creature Labs and Dundrum Bay Oyster Fishery illustrate two different ways in which this model can be interpreted. Chris McKee's willingness to take calculated risks has placed Creature Labs at the forefront of computer game development. At Dundrum Bay, necessity has been the mother of invention for Robert Graham, but his in-depth understanding of the operational techniques conditioning his stretch has helped ensure his success.

One thing McKee and Graham share is a coherent vision of what constitutes business success. The proceeds of their creative capacities could easily have been squandered had it not been for their focus on achieving measurable objectives in realistic timescales. The desired ultimate aims of stretching performance should, therefore, be clear from the outset, with key tasks detailed and milestones agreed.

But what is the best means of capturing this information? The answer lies partly in business and financial planning. Creative thinking is great but ideas need to emerge in a commercially viable form. The methodical approaches of bankers and accountants might appear incompatible with the spontaneity upon which innovators thrive but when all parties share a common aim this unlikely alliance can prove fruitful.

Besides encapsulating the objectives, financial estimates and forecasts associated with a prospective innovation, a well-written business plan can be used to summarise the state of the business as a basis for judging the need to secure extra funding. The plan should then be tailored for prospective investors, identifying what is exciting and different about the creative idea and demonstrating that they stand a good chance of being paid back or getting a good return on their investment.

The supplementary benefit of business planning is the crystallisation of objectives and timescales. McKee and Graham have both taken into consideration the resources and interim deadlines required for their innovative thinking to pay dividends. Where organisations are complex in their structure, internal dependencies are also highlighted.

What the business plan must not become is a triumph of style over substance. Quality content is essential and should be geared throughout towards reducing risk. Where innovation drives development, lack of demand for the finished article poses the biggest threat to success. Creature Labs' profitability shielded them against this, whilst Graham's rotor was designed to solve his own operational problems, diluting the threat of wider marketplace failure. Market research illustrating a demand for the product is the best alternative for innovators seeking to persuade investors that their plans are viable.

The reliable delivery of products to customers is another hazardous area that business planners must address. With Creature Labs already established as a market player, McKee had the human resources needed to produce the game and the customer base that was likely to purchase it. Graham faces a more challenging task because growing oysters is a world away from manufacturing and distributing rotors. Should he choose the manufacturing route, the production process would need to be smooth from the outset to keep customers satisfied. Partnership agreements with specialist manufacturers would be a sensible move, although inevitably the bounties will have to be shared.

A business plan should act also as a bulwark against unrealistic financial estimates. Financial contingency planning is essential in funding any major project, and the business needs to be resilient enough to absorb unanticipated expenses. Those looking to challenge this view need look only at the costs of the Euro Disney, Channel Tunnel and Millennium Dome projects to guide their thinking.

Including the above points in a succinct business plan should put any business in a position to stretch itself effectively. But success will depend ultimately upon communication of the innovator's vision to those staff whose support is needed to realise his or her dream. Employees should be kept informed throughout the change process and their feedback sought to create a feeling of common ownership of the project. Besides helping owners like McKee retain key team members, the risk of rivals successfully 'poaching' important keepers of knowledge is greatly reduced. Even so, both Graham and McKee should protect their intellectual property rights so would be well advised to make sure their ideas are patented.

Whatever the outcome, innovation is the lifeblood of development and both businesses should be saluted for their achievements in this area. Business and financial planning might not seem as romantic as creating new products but if done properly, it can provide the framework required to channel the flow of ideas into something likely to achieve the commercial success they deserve.

1

Unlike people, characters in computer games do not adapt and learn to take advantage of their opponents' weaknesses.

Creature Labs recognised the potential of applying scientific research to artificial life and applying it to computer games. This enabled the Cambridge-based company to create innovative 'living games' that have captivated a global audience and transformed the company's profitability.

'There is something about biological intelligence that is way ahead of anything we can engineer at the moment,' says Chris McKee, CEO at Creature Labs, a computer games company that employs 70 people.

In the early 90s, it focused on developing conventional computer games. 'But we didn't want to make "me too" games like everybody else,' says Chris McKee, 'so we spent a few years thinking what we could do to differentiate ourselves from other game developers.' The response of the company's chief technologist, Stephen Grand, appeared wacky at first: the only common denominator of all intelligent systems on earth is that they are alive. Creature Labs backed his hunch and demonstrated its willingness to take risks by giving him free licence to explore the concept for two years. Its commitment to innovation brought valuable returns.

At the end of this two-year period of initial development, Stephen Grand had created a set of chicken-like creatures that operated, walked and fed by biological simulation. Was there scope for a new style of game? Creature Labs believed so. After a further two years, Creatures was ready. Uniquely, it was capable of running several discrete neural networks, each one representing a Creature's adaptive 'brain', in real time on a standard PC. The principles were highly unusual. Participants do not win when they play. Instead they learn from their mistakes when losing and play more effectively next time as a result.

Creature Labs took its innovative game idea to Warner Interactive, a new arrival in the industry. After some persuasion, a non-refundable advance of £1m was secured, which represents 200,000 units of sales. In fact, half a million copies of the original Creatures have been sold so far, and in addition to being selected by the Design Council for Millennium Product status, it has ranked among the top ten selling PC games.

Chris McKee warns against compromises. 'Don't change your original hunch. Don't lose your original belief,' he says. 'If you get persuaded that half this way, half another is safer, you'll never actually do it, because you won't have the same passion. But be savvy enough to talk to people outside. Although technology is never wrong, the market may just not be ready for it yet.'

The development process for Creatures took more than four years. 'We didn't put any harness on Steven and his team, saying you must produce something marketable,' says Chris McKee. 'Since we made profits in other parts of the business, we were able to accommodate free-ranging ideas.'

'With our particular technology we took a very clear and intentional decision not to modify it to suit any particular customer,' says Chris McKee. 'We realised it would have potentially generic appeal. This was a high stakes approach because it was one that said we would either be highly successful or fail completely.'

case 2

hatching an idea

'If there's a problem, you don't just fix it; you solve it.'

Robert Graham has always relied on intuition. After reading an article on oyster farming, he believed so strongly that the popularity of oysters would increase with the trend towards healthy eating, that he and his brother set up Dundrum Bay Oyster Fishery. A few years on, Graham gave up a 17-year career, a home and a guaranteed income in order to buy out his brother and manage the business, which was only just breaking even. He was well aware that the operation needed to increase substantially in scale but retain the existing standards of quality.

As a one-man-band, this was difficult to achieve; oyster containers need to be turned regularly and Robert was struggling to do so on his own. He had insufficient resources to employ labour or to grow the business. For the next few years, Robert dedicated himself to the process of innovation, brainstorming and poring over designs whenever he could find the time.

Eventually, with the help of a design consultant and some grants, Robert developed the Rotor; a device using a float and the tide to regularly turn containers. Not only has the Rotor won the status of a Millennium Product, it has also raised his company profile, significantly increased productivity and quality, and made operations more economical.

Rotors are now producing some, but not all, of the Fishery's oysters. As a cost-driven and design-driven product, the Rotor is simple, robust and economic enough to revolutionise the oyster industry. The more frequently an oyster is turned, the better the quality of the product, but the process is labour-intensive and costly. Industry best practice is for oysters to be turned once a month, but Robert's innovation enables the oysters to be turned automatically with every tide: twice a day!

The Fishery has always been small; Robert and his brother managed it part-time for eight years before Robert went solo. Even now, he employs only four people. Many would have given up on innovation years earlier. By 'setting aside daily creative time to devote to the problem, much like a writer does', Graham invented the Rotor.

His ethos is one of design excellence. 'If there is a problem, you don't just fix it; you solve it.' The 'designing out' of problems was necessitated by the shortage of time, employees and financial resources in the company's early years. Initial difficulties were overcome by developing a cylindrical oyster bag. Then Robert, who had time to turn the bags only twice a year instead of the best practice of once a month, hit on the idea of using tides and a float. Finally, he brainstormed on how best to put these ideas into practice. 'There are several tiers to the design process,' says Robert. 'Once you have the solution to a problem, there is another tier of problems underneath.'

The results of the first year's trials of the Rotor were spectacular in terms of the quality and yield of oysters produced. However, lessons were learned; whilst the concept was well founded, the practicalities (and, in particular, the costs) of making the Rotor needed re-addressing. Robert realised that he would need outside help, brought in design consultants and by working together they were able to resolve the difficulties.

When Robert felt that he needed some external expertise, he attended a Northern Ireland Office design seminar, at which he met and selected Isis Design. His choice was made mainly on the basis that Isis had the right expertise but it also relied on Robert's sense that he would get on well with the consultant involved. He was right; due to a close personal and professional fit, the partnership has been successful. He is also working with Queen's University's Plastics Development Centre to make the design more economic.

Graham was careful to minimise risks within the design process. The consultants, Isis Design, were given an explicitly detailed brief. The key measure of feasibility was the cost per oyster per season. Robert feels that it helped to start from such a tight brief: 'Costs rose, but unit capacity also rose from 100,000 seeds to 250,000.'

In 1993, grant aid and bank loans enabled Graham to invest in new premises and to take on two full-time staff. This was a major step in allowing him to concentrate on raising quality standards with the resources that he had and, ultimately, to develop the Rotor. A SMART award helped him to finance the product's development by covering 75 per cent of the £50,000 costs of producing the first working prototype.

Robert worked in close partnership with the consultants, with brainstorming sessions and ongoing dialogue. The consultants took very careful account of Robert's needs to minimise the costs of manufacturing, assembly and materials. Their close teamwork led initially to the creation of a booklet of ideas for each of the components of the rotor. This provided the basis for the production of a fibreglass prototype and a successful trials programme.

Robert applied for a patent for the Rotor before the prototype was tested. He also made Isis sign confidentiality agreements. Despite the advice he received on IPR, he still found the patent system to be a 'nightmare'; the costs were crippling. Graham has reservations about the benefits received from his patent, given the high costs and the lack of cheap, good advice; 'It's one thing taking out a patent,' says Graham, 'it's another thing protecting it. I decided to go for it, but I'm very sceptical of the benefit I will receive in the long-run.'

Graham's objective is not only to run a profitable oyster farm. His close under-standing of the market has revealed a need amongst fellow oyster farmers for a means to regularly turn oysters in a manner that was less labour intensive and costly. He believed his Rotor device satisfied this need. His investment in building an understanding of customer needs paid off in 1997 when he took the Rotor to an aquaculture exhibition in Scotland and it attracted great interest. The main area of customer resistance concerns the high capital cost of the machine (£2,500), despite the fact that each one would pay for itself in one to four years. Robert is therefore considering leasing the machines or entering a partnership to market them.

1.4 where people come first

'capitalism has run its course and now it is time for an economy based on people power to take over'

Interview with Professor Alec Reed CBE, Founder and Chairman, Reed Executive plc

contributor profile

alec reed

founder & chairman

Professor Alec Reed CBE is the Founder and Chairman of Reed Executive plc, one of the largest recruitment solution providers in the country. Reed Plc has an impressive high street presence with over 280 branches and is the market leader in Internet recruitment with its site, www.reed.co.uk receiving upwards of nine million page impressions every month.

Alec is a Fellow of three leading institutes representing Accountancy, Personnel and Marketing and is an enthusiastic, innovative and entertaining contributor to the development of each profession.

With an abiding interest in education, Alec set up the Reed Business School and founded The Academy of Enterprise, a non-profit initiative that promotes enterprising behaviour throughout the United Kingdom. As Professor of Innovation at Royal Holloway, University of London, he has created a unique course for undergraduates concentrating on Leadership, Innovation and Enterprise Studies (LIES). This has been developed at London Guildhall University, where Alec is a visiting Professor of Enterprise, into a BA in Business Enterprise.

With over four decades of business experience and as a respected academic in the field of Innovative Practice, Alec has recently published a book on Innovation in Human Resource Management with the Chartered Institute of Personnel and Development.

1.5 learning and development

'one good way of thinking about learning organisations is to understand the nature of the fitness landscape on which they are operating, and to try to understand what steps can be taken to smooth that landscape'

**Vince Darley PhD,
CEO, EuroBios UK Ltd**

contributor profile

vince darley

**president
and CEO**

Vince Darley is currently CEO of EuroBios UK Ltd where he manages an excellent team of scientists and developers, delivering highly optimised solutions to a wide variety of routing, distribution problems, and modelling operational risk and adaptive ecosystems.

Vince started his academic life with a mathematics degree from Cambridge University, then moved through theoretical physics and computer science degrees before being seduced into complexity by the Santa Fe Institute. His doctoral dissertation was concerned with understanding the natural dynamics of large systems of autonomous, optimising agents, and how to design local interactions in order to bring about particular global goals.

At Bios Group, he applied these insights to understanding the behaviour of stock markets, service organisations and supply chains using agent-based simulation, and developing more robust, dynamic scheduling algorithms.

learning and
development

What is a learning organisation? Perhaps the first interesting results in this area came from observations made during the Second World War on the construction of bombers. An analysis of the efficiency of production showed that with each *doubling* of the number of bombers constructed, the cost of production decreased by a constant fraction. This 'learning curve' or 'experience curve' phenomenon has since been observed in an enormous variety of systems. When a scientist sees a regularity like this: something robust, reproducible and precise, they look for an underlying law and underlying reasons for that law. Current research suggests that this regularity can be explained through the study of search processes on 'fitness landscapes'. A fitness landscape is a way of representing the impact of small changes to a process. Each point on a landscape corresponds to one mode of operation, and the 'height' of each point represents its efficiency ('fitness'). Therefore, one wishes to be on as high a peak as possible for maximum efficiency. Through production, one learns to make small, incremental changes to a process, each of which corresponds to a single step in the landscape. Through repeated steps, one can gradually move 'uphill' to higher and higher peaks. Clearly, moving uphill in a very smooth landscape is quite easy, but doing so in a rugged landscape with large valleys, crevasses and cliffs in much more difficult. By applying the 'NK' family of fitness landscapes invented by Professor Stu Kauffman of the Bios Group, we can show how the ruggedness of the landscape representing a particular process is directly responsible for the shallowness of the learning curve. On a smooth landscape you learn quickly (the constant learning fraction is large), but on a rugged landscape you learn slowly (the constant learning fraction is small). However, the learning always follows the same mathematical form.

What this teaches us is that one good way of thinking about learning organisations is to understand the nature of the fitness landscape on which they are operating, and to try to understand what steps can be taken to smooth that landscape. Not surprisingly the more that different teams in the organisation work separately, in their own silos, with limited communication and significant potential for conflicting constraints between the teams, the more rugged the landscape is. This may not be surprising, but the fact that new scientific tools and methods are available to understand and optimise the landscape, is really quite new and exciting. For example, the Bios Group has applied these techniques to help Boeing understand how to make the airplane design process more efficient.

Learning curves naturally seem to apply to a very 'old world' picture of a monolithic corporation making large physical goods. They aren't actually limited to just that picture, but it is true that the newer, more distributed and dynamic 'ecosystem' in which most companies operate today, is different. Can the scientific perspective provide insights and new ways of thinking about learning organisations in such new environments?

Our experience at the Bios Group has been that it can. I've isolated a few key points that might be useful to the reader:

'Patches' – one cannot optimise a large organisation from the head-office. But also, one can't allow each individual employee to optimise their own behaviour without paying attention to anyone else (10,000 silos!), since each change made by one employee would ripple through the organisation resulting in chaos. Research in complexity science has shown that for each organisation there is an optimal 'patch size' which is, loosely speaking, the nearby portion of the organisation on which each individual should examine the consequences when they make their decisions. Too small a patch size and one has chaos, too large a patch size and the organisation is frozen in stasis.

'Look at the ecosystem' – the whole *is* more than the sum of the parts, and nowhere is this more evident than in the business world. While economists still argue about why firms exist, the business world gets on with forming, merging, spinning off (and, sometimes, closing) firms in a complex ecosystem of goods and services. Each firm itself is a complex, adaptive web of interactions between people, information, products, transport, etc. There is a growing belief that successful companies must understand how to capture the power of invention and learning inherent in the external ecosystem, and, where possible, to bring that power *inside* the organisation. It will mean a more fluid organisation, perhaps more chaotic even, but the result will be more nimble, adaptive and successful.

'Phase transitions' – small modifications can result in big changes (both good and bad). We are all familiar with the way in which water, as it is cooled from say 40° to 30° to 20° to 10° to 1°, appears almost completely unchanged until we reach 0° Celsius and then, all of a sudden, it 'freezes' and exhibits a completely different behaviour.

Actually, if the water is very pure, we can actually cool it below 0° without it freezing. But then, if we introduce a grain of dust (or there is a blemish in the side of the ice-tray), suddenly the water will freeze. Look at the way our national rail network operates, or the way our underground network performs. It used to be reasonably reliable, but then over the years, as investment was reduced, as maintenance was (arguably) delayed, as the number of trains running was increased, we have now reached exactly such a fragile state where, if everything goes perfectly smoothly and according to plan, the system keeps flowing ('water'), but then introduce a slight hiccup and a cascade of reactions and inter-reactions causes the system to transform almost instantly into a very poor state, with huge delays ('ice'). Systems pushed close to, or even beyond their phase transitions are very, very delicate.

At the Bios Group, one of our projects involved examining a large factory operated by Unilever in the UK. It turned out that this factory was showing exactly this kind of 'post-phase transition' behaviour that we are all so (unfortunately) accustomed to today. So, while an organisation learns and explores and adapts, it needs to monitor itself and its processes and understand when it is getting dangerously close to such phase transitions. This is all the more important in the business world, since a common equivalent of the 'temperature' is the 'costs' the business incurs – excessive cost cutting is one sure way of running into your nearest phase transition.

Of course, phase transitions are often negative things, but they can equally be viewed as very positive things. If you can identify that you are on the wrong side of one, a small change or small investment could cause dramatic improvements – let's hope this is also true of the rail networks (of course, where and when and how that investment should occur is key to getting to the other side of the transition).

'Emergence' – phase transitions are just one example of what are now known as emergent phenomena. The learning organisation must always be on the lookout for these. Continuous incremental improvements should, of course, be the underlying learning strategy for most companies, but in addition there must be a mechanism for generating the step-changes. The study of fitness landscapes and the trade-offs between exploration and exploitation has shown that significant improvements are very unlikely to be reached by a series of small, positive changes. 'You can't get there from here' might be a good motto. Step-changes usually require a combination of numerous small modifications in many parts of the organisation, and therefore, to be successful, must be carefully synchronized and well-understood events. To get from one stable operating state to a new one requires this. This is especially true given that each state is usually quite robust and stable, even without taking into account people's resistance to change. If we don't make the right combination of modifications, almost surely the organisation will slip back into the old operating methods.

You can treat these key points either as simple analogies and metaphors to aid in adding new thinking to old problems, or you can take them one step further. By careful, rigorous use of scientific tools and methods and agent-based simulations models (ABMs), all of these phenomena have been shown to be very real and present in the business world. With appropriate modelling and use of these techniques, one can really understand and predict, quantitatively, the outcome of certain kinds of behaviour changes and investments in organisations for the first time.

case 1

execute with extreme speed

In the computer-games business, you have to work quickly.

'When you've decided what to do, make sure you've got an organisation that can execute with extreme speed,' says Chris McKee, chief executive of Creature Labs in Cambridge, which has created an innovative 'living game' using biological intelligence.

'Don't take a year to execute and a month to think,' he says. 'Take a year to think and a month to execute. That is a formula for very successful companies.'

Creature Labs is not aiming to become a large company. Chris McKee believes that by working smarter they could generate five times current revenue without employing more than 80 people. If staff numbers were to grow beyond this point, he believes, there would be a danger of losing culture, passion and focus. Nonetheless some procedures are being put in place to prevent areas of minor conflict that have arisen as the company has grown.

Most people at Creature Labs are involved in building games using current technology. A small group has been at work inventing the next generation of technology.

Creature Labs 'grows' people by mentoring, giving them opportunities and offering support. Recruitment is by contacts rather than advertisements. It would be unusual for them to take on anybody without the requisite technical skills. Existing team members participate in the recruitment process.

Funds are allocated for training. If an employee has an interest that could benefit the company – such as one employee who wanted to take an Open University degree – then Creature Labs provides 100 per cent financial support. Performance reviews are held every six months with bonuses of up to 10 per cent of salary allocated for exceptional performance.

The office is structured to enable team working, discussion and cooperation. Teams divide up. A producer is in charge of each overall project, but there is the potential for project champions to lead. It is a loose, adaptable structure. 'You see things, you hear things, you get involved by default,' says Chris McKee. New employees are easily accepted and existing team members get involved in the interview process.

Chris McKee believes innovative people spark creative ideas among their colleagues. 'If you are a creative business, you really need people together all the time,' he says. 'A lot of good ideas come totally unplanned in a serendipitous way. It's very hard to do that remotely.' Much of the creative spark is cultivated in Henry's bar beneath the office. Cross-fertilisation of ideas combines with peer pressure to produce an innovative culture where it is common for people to work through the night and at weekends. But everyone has to be at work by 10am 'in order that a buzz is created'.

It gives participants the chance to think more laterally, corresponding with Thames Water's efforts to recognise and reward creativity in all its forms.

case 2

games bring gains

Thames Water's Network Challenge captures the interest of schoolchildren in complex engineering tasks. It is also pioneering a new approach to innovation within the company. The Network Challenge is an educational tool originally designed for schools by Thames Water. In about three hours, groups are challenged to construct a water system, not on paper but on a three-dimensional model.

The experience is designed to be both fun and interesting, giving children an insight into what engineering involves. It is a formula that works with more grown-up audiences as well. As a result of playing, new recruits, non-engineering staff and customers have a better appreciation of the complexities involved in water projects. The Network Challenge has been run, for instance, at a corporate finance away day and at a call centre in Swindon to give operators a better idea of what customers were complaining about.

The challenge is to build a pipeline to supply five new customers in a variety of terrains. The kit comes with a large, colour printed board and a set of pipes, connectors, flow meters and water containers. Land areas have point values to indicate how expensive they are to build across.

Up to eight teams of five players compete. Their goal is to build a network on time and to budget. Facilitators from Thames Water guide the participants through planning and procurement, construction and commissioning phases, providing a taste of the real – and often unexpected – hurdles that are encountered.

The Network Challenge is a multidisciplinary effort, calling on individuals in corporate affairs, engineering and education units. It is not just about cost. It gives participants the chance to think more laterally, corresponding with Thames Water's efforts to recognise and reward creativity in all its forms. The Network Challenge was an early example of the benefits of this approach and it is held up as an example of how employees can work together in cross-functional teams.

The original idea for the Network Challenge was an amalgam of three to four ideas from graduates, who had been asked to think of projects that were practical and enjoyable while still challenging. New facilitators are being trained to take the Network Challenge to a wider audience and the original team members are slowly handing over the future development of the project to a new generation of young Thames Water engineers.

The Network Challenge is copyrighted. As long as it receives recognition, Thames Water does not mind it being copied. In fact, it would like it to be adopted by other professions. There has been some interest in the concept from National Grid and the Network Challenge is now being played in China, Thailand and the USA, where Elizabeth Town Water have bought six of the role-play games.

case 3

a leading light

Select Software has come up with a potentially world-beating product after an accidental contact with the lighting industry.

Select, which is a small family business run by David Aarons and his father, was working on a timing device for swimming competitions. One of its suppliers was having problems with the neon lighting for its control devices. Select produced its own solution, which prompted Aarons to take a wider look at the lighting market. He learnt quickly.

The 'dimmable ballast' which Select subsequently developed won the International Total Lighting Award in 2001 and was awarded the status of a Millennium Product by the Design Council. It is designed to reduce energy, while allowing more effective lighting. This efficiency means that the ballast runs at a much lower temperature, cutting back on levels of cooling.

Although a small company, Aarons believes that keeping up to date with technical know-how and commercial experience is of vital importance in staying ahead of the field. Despite having no formal qualifications, David has learned through experience. He now keeps up to date with new technology through suppliers, using what is most appropriate for the job in hand.

Aarons believes in carrying out sound research of the product requirement. This is based on discussions with people in the market and with the use of his technical knowledge, asking searching questions on what might be required. Rich sources of information such as service records, complaints and help lines are used to find out the needs of the market. For the dimmable ballast, Select used its own market knowledge but also brought in specialist help when required.

There is no arrogance about knowing best at the company. Select accepts its limitations and recognises the importance of building a strong team to complement in-house resources when they are not adequate. When you have reached the limit of your potential 'employ people who know more than you' and go for the best available, says Aarons. To help with the dimmable ballast Arons brought in top patent advice, a large accountancy firm to advise on financing the new venture and a non-executive chairman with experience in a major lighting company to lead the project.

Select does not intend to market the dimmable ballast itself; this is primarily because they believe it is too big for them. They will instead licence it to major firms in the lighting market. The company believes that income from royalties could be massive.

2 Create

2.1 teams and networks

'innovation is the emergent quality of your teams and networks'

**Garrick Jones,
Executive Consultant,
Cap Gemini Ernst &
Young's innovateUK
centre**

contributor profile

garrick jones

executive consultant

Garrick is an Executive Consultant and leads the team at Cap Gemini Ernst & Young's innovateUK in Soho, London. He has ten years' experience in Transformation Consulting and has worked at the heart of Cap Gemini Ernst & Young's Acceleration centres around the globe since their inception. In this capacity he has worked with many of the Fortune 100 bringing innovation and transformation to their most complex programmes.

Garrick's work at innovateUK is linked with the Centre for Business Innovation in Boston. Our Innovation and Solution Centre's work closely with all of Cap Gemini Ernst & Young as they assist companies with innovating products, supply chains, strategies and adaptive technologies. Garrick studied at Oxford University.

teams and networks

Competitive advantage can be described as the ability to learn, innovate or continuously reposition with respect to the competition. Enabling a culture of continuous learning and innovation requires processes in place, which are as valuable to the organisation as those of cash flow and budgeting. Creating such a culture requires continual sponsorship from the most senior levels. One of the best ways of doing so is to promote organisational learning and enabling employees through networks and partnerships which provide new ideas, new skills and new capabilities.

Radical innovation – defined as the creation of entirely new sets of performance features: at least a five-fold improvement in known performance features and a significant (30 per cent) reduction in costs – is best facilitated through the use of network hubs or centres of excellence where organisations are able to converge the ideas of multiple networks around particular solutions.

As we move to a networked economy the concept of the linear supply chain has transformed into that of the non-linear value web. Successful organisations are able to identify the members of their value web and create opportunities where all their resources can be brought to bear on existing products and new ideas.

The companies that are most successful at mobilising their value webs are those that enable frequent interactions with their networks around specific projects.

How do you set up the preconditions to ensure you make the most of your networks? How do you ensure your innovation teams will be successful?

Innovation is the emergent quality of your teams and networks. Innovation teams and networks are the source of proven new ideas in your organisation. To enable the emergent quality of their work to be innovative they need to be:

- *Autonomous*
  Teams need to be given the space and freedom to explore ideas that don't necessarily follow the organisational norms. Some organisations set up 'skunk works', which exist to develop new products and ideas, test and pilot them and then transfer them to the organisation. They stand alone but their members contain people from across the breadth and depth of the organisation who rotate through, bringing both the understanding of the business, and creating support for the new projects.

- *Configured with the best members for the task*
  Multi-disciplinary teams are hard to manage, but they are vital for the creative dynamic that sparks new ideas. Hire mavericks and provide opportunities for teams to work with ideas generated from the entire organisation.

- *Connected to learning networks*
  Build networks with academics, research, leading thinking, conferences, universities, college projects and enthusiasts. There is a tremendous amount of thinking going on out there, tap into it. Allow your new products to be the test beds for the new ideas being generated in the broader society.

- *Connected to your customers*
  Your customers know what they need and how they use your products. They may be doing so in ways you never anticipated. Some of the most innovative products are now providing the tools to their customers to transform the products themselves. For example, Reason 2.0 – one of the leading music generation products – enables customers to create new modules and then acts as a broker, creating the links between their customers to share the new tools.

- *Connected to your value web*
  The opinions of your clients, employees, suppliers, customers and learning networks enable the ideas generated to be greater than those existing within the organisation. Encourage osmosis of ideas. In addition to generating ideas, you also begin to mobilise the users of the products, creating the buzz around the new products long before they are launched, and creating an influential user community in the process. ITS Technology demonstrate how working closely with your clients at the design stage of production innovation ensures you have an engaged set of clients before you enter production.

- *Skilled in disciplines associated with innovation*
  Innovation teams need skills in accessing creative thinking, creating artefacts that represent their ideas, and ideas logging so that databases are created which feed the ideas generation process. Clear understanding of where their ideas are in the process from generation, testing, realisation and production is also important.

- *Incentivised*

  Although teams need to be autonomous, it is important that the members of the teams feel rewarded for the work they are doing. The design process honours failure – high-volume, low-risk failure. The iteration process leads to better outcomes. Incentives link the outcomes with the broad requirements of the organisations, and enable teams to maintain focus even when things may not be going well. Reward failure as well as success. Failure is where most learning takes place.

- *Measured*

  Measuring the success of teams against understood criteria, established clearly at the start, provides the security that teams tasked with thinking widely and differently need, as well as providing benchmarks against which organisations can measure the success of their innovation programmes. The creation of milestones in project lifecycles, as well as the production of artefacts, ie documentation, communication materials, proto-types and other media, provide visible and tangible measurements. Innovation is not an intangible art.

  TR Europe's success with the Aluminium Bumper Nut was directly linked to the quality of the partnership it pursued with Hydro Raufoss.

- *Sponsored*

  It takes powerful leadership to create the freedom and context in which innovative thinking can flourish. Sponsors understand the end game and provide encouragement, guidance and support when things may not be going well. Sponsors also ensure that the best ideas are shepherded through the organisational processes required to get them into manufacturing or operations. They understand what the end game is all about, although they may not know what the final outcome is. How many great initiatives fail at the final fence because they haven't been supported powerfully?

  A&R Cambridge Ltd developed the successful Arcam Alpha10 by partnering with sources of new technology not available in the firm. Their success not only demonstrated the value of partnering to source new skills and technology to create entirely new products, but would not have been achieved without the innovation sponsorship provided by John Dawson.

Engaging a broad community around product innovation greatly increases the chances of success when your product comes to market. Involving your entire Value Web in the creation, design, testing and operation of your products creates a community that is emotionally engaged. Networks are powerful sources of new ideas. Manage your networks strategically. Identify the key members and involve them. Create opportunities for frequent interaction between your networks and your innovation teams, so that your organisation taps into the rich source of ideas and research that exist around you.

case 1

The team was unusual, with individuals based in different countries and involving both a customer and a supplier.

trusted relationships

TR Europe uses teams of five to six people in product development as a matter of course. In the case of the Aluminium Bumper Nut developed for BMW, Colin Mileham as manufacturing manager headed a larger core team that incorporated engineering, sales, design and production skills. The team was unusual, with individuals based in different countries and involving both a customer and a supplier. Particularly important was the fact that both engineers and manufacturers were involved – with the former taking the customer's view and the latter considering implementation.

The idea for the Aluminium Bumper Nut was fundamentally customer driven. BMW approached Hydro Raufoss, their Norwegian bumper supplier, asking for a better looking, single-operation recyclable fastener than the one currently available on the market. The recyclability factor meant that the fastener had to be made from aluminium – an approach that had never been taken before. The aluminium would need to be sufficiently harder than the bumper itself. BMW were also insisting the fastener should not be in the traditional hexagonal shape (which needs three hands to fasten) and there should be no compromise in mechanical efficiency.

Hydro Raufoss approached TR Europe with the challenge, which Mileham was keen to take up. Mileham scanned trade magazines to identify possible materials and a leading-edge supplier. The team – which expanded to include BMW and Raufoss – looked at various sectors, eventually finding a solution in an aerospace product. 'In the whole of this process,' says Mileham; 'more ideas came from the shop floor than anywhere else.'

The company previously ran an internal campaign called 'Make a Difference' (MAD) but it was little more than a suggestion box scheme. The company found that this technique led to an influx of 'silly ideas' and now employs a 'semi-driven' policy where suggestions such as this are sought during regular team meetings.

Much of TR Europe's success depends on the partnership that it actively pursued with Hydro Raufoss. Issues relating to quality, intellectual property and recycling are all developed from a basis of trust and a good relationship. Likewise, such a relationship acts as an enabler for the exchange of technical information. Of all their practices, Mileham believes the openness in relationships to be key to their success.

As well as a fundamentally customer-driven design process in which the customer is part of the design team, a dedication to quality lies at the heart of the company's customer focus. TR Europe has achieved a number of quality standards and is now aiming for the QS9000, which it feels will be the norm in the future. The company was the seventh to achieve BS5750 in 1982. 'TR stands for Total Reliability,' says Mileham.

In an attempt to avoid repeating mistakes, the company uses Failure Mode Effect Analyses (FMEAs), to predict scenarios and achieve quality. Benchmarking is also used to evaluate performance against targets and competitors, and total quality management is also embraced. TR Europe carries out regular re-evaluations of its manufacturing and engineering processes – even where a system is working well – to identify areas that might be improved.

Though the team represented a costly overhead, Linear Drives Limited has maintained it 'even in hard times'.

case 2

through thick and thin

Linear Drives Limited is a small engineering firm founded by a seven-strong team that obtained start-up backing from a 'business angel'. The company now employs 46 people and specialises in making motors for industrial automation. It has won a Design Council Millennium Product award for its development of a new form of electric 'linear' motor, branded 'ThrustTube'.

Instead of relying on conventional rotary mechanisms, which are vulnerable to wear, its new straight-line motor offers consistent control without the same level of damage caused by use. The new technology is adaptable to many industries, from food to bio-tech, but Linear Drives Limited is currently selling principally to the electronics industry, where its motors are used to drive automated production lines.

Staff at Linear Drives Limited were never in any doubt as to the level of commitment within the company to innovation. It was clear that the managing director, Bill Luckin, himself was driving development of the new linear motor forward from the top. He refers to the technology as a 'great simple idea – a combination of speed, accuracy and reliability'.

Linear Drives Limited invested in a team of seven in-house development engineers, which represents a large commitment for a small firm. The team was there not only to develop ideas but also to make sure the technology was translated into commercially viable products. Though the team represented a costly overhead, Linear Drives Limited has maintained it 'even in hard times'.

Luckin and his partners have taken care to build a close, informal and friendly work culture. Everyone from managing director to receptionist is encouraged to put forward ideas and to approach life as a continual learning experience. The company's openness to learning is underlined by its willingness to seek and accept help wherever this is offered.

From the outset, the development of the new motor was driven forward by Luckin, partner Scott Pleva and the seven-strong team of development engineers. But more than this, the company has attempted to implement team working all the way through its operations. Its shop floor is divided into teams run by team leaders and a working day begins with brief team meetings at which everyone is encouraged to have their say. A two-hour meeting for all managers every Monday addresses general issues from staffing to project management and finance. These regular meetings are designed to encourage interaction to iron out problems and enhance motivation.

case 3

inertia is the enemy

'They are constantly innovating, in terms of identifying ways of improving the product. They usually tend to tell me after they have done it, rather than beforehand.' A technical discovery at UMIST has developed into a process that could take the place of X-rays for many commercial applications. Conventional tomography, the process of applying sensors around an object to determine what is inside, is expensive. A CAT system used in a hospital costs £300,000. The cost of ITS's rapid process tomography brings it in range for industry and won Millennium Product status from the Design Council. The system can be set up to run off a PC for £40,000. ITS's approach is to sell it as part of a service package, rather than as a standalone product. Its engineers identify clients' problems and run a feasibility study to work out the implementation of the technology.

In ITS's office in the centre of Manchester, Ken Primrose, the chief executive, sits with his team of nine with no obvious distinction drawn between personnel. His view is that inertia is the main competition and employees are encouraged to come up with ideas. 'They are constantly innovating, in terms of identifying ways of improving the product. They usually tend to tell me after they have done it, rather than beforehand.'

The technology is seen as having great potential and the team wants to be part of it. 'It is very exciting to be part of a new technology.' Success is celebrated and all employees are aware of what is happening in the company and where it is heading. Formal appraisals are carried out, though reward is not directly associated with this.

ITS is more a small team than a company, motivated by the excitement of working with a leading edge technology. Employees are set personal objectives. They network mainly with academics through sponsored research and conferences. As a group, the ITS team has to yet to face any failures. 'The only downer is the lag between someone saying they want a system and actually buying one.'

There are two types of customers for ITS's instrumentation, research bodies and multi-national companies. The key for ITS is to work with clients at an early stage of development. 'Once the process has been licensed people are pretty resistant to doing new things. If we are going to get involved in anything it has to be at the manufacturing development stage.'

The company needs to work very closely with the customer as it is a consultant/client relationship and feedback is essential in order to deliver the necessary service. They do not use distributors because the aim is to achieve a high level of understanding and service.

ITS's hardware is standard, but the sensor is bespoke for each application. 'There is direct interaction on site, where we can see what their frustrations are, and how the instrument might be improved to meet their needs,' says Primrose.

Beyond the board Gary Meadows knew he had other stakeholders who were likely to perceive this new way of working as threatening. He anticipated this by seeking the buy-in of employees. He turned the threat into an opportunity by actively selling how the new system would help them re-skill rather than de-skill. Within 18 months employees were champions of the cause. The project made sure that it connected across the internal boundaries of the AA. It worked not only with call handlers but also those that would be out in the field living with the decisions made by the new system, the roadside staff. Gary Meadows and the system's original designer, Nigel Brown were given great freedom to pursue the project. The AA committed to the programme even though it didn't have all the knowledge it needed to be sure of success. The AA had created a series of real options that they could exercise or not as they faced each milestone. With each milestone their knowledge and understanding of whether this new system was going to work improved. As a result, the UK's dominant provider of roadside assistance has successfully led the development of an innovative new call handling and dispatch system that will mean it can remain competitive even as new entrants try in vain to muscle in on its cherished position.

JCB, like the AA, is recognised as an industry leader. It's a dominant force in shaping and leading the market for agricultural and construction vehicles. When Sir Anthony Bamford, the company chairman, made the decision to attack the market for forklift trucks it acted like a new entrant. If JCB were to be successful in this highly competitive market, it would have to use innovation as a disruptive force to undermine the considerable advantages of the market's incumbent players. JCB had a concept product under development. Even though they didn't have all the information they needed about the concept and the market JCB decided, in a bold move, to commit to launching it at a major trade show. JCB knew that if it simply aped its competitors it would be crushed. It knew that it would need a new and innovative approach to the market. The project team responsible for this new product, like the project at the AA, were encouraged to reject the familiar in favour of the unfamiliar. The team was able to explore new ideas in an informal and unstructured way. The team connected across the boundaries of JCB forming a multi-disciplinary team of engineers, designers, dealer representatives and production and service staff. The team worked hard to promote its radical vision for JCB's forklift product to ensure that the organisation shared the enthusiasm for the opportunity that this innovative entry into the forklift truck market represented.

1

Dawson believes that you have to drive what you want back down the organisation. Britain's leading supplier of hi-fi amplifiers and CD players began as a sideline at Cambridge University. John Dawson found that designing and selling hi-fi equipment for student discos was more interesting than his doctoral thesis. A quarter of a century later, A&R's hi-fi separates now outsell Sony and Technics in the UK. The company owes its position to a combination of astute marketing, innovative engineering and, above all, the unstoppable enthusiasm and enterprise of John Dawson.

He is a confident and enthusiastic managing director, who is not slow to display his product mastery in a highly competitive sector. He still holds to the vision that inspired the launch of the company's first hi-fi amplifier in 1976: both the quality and cost of sound equipment in the mass market can be significantly improved by utilising the latest developments in technology for components. 'Producing an innovative product,' he says, 'gives you more freedom within which to set price.'

When a UK competitor threatened to lift customer expectations to a higher level in 1996 with a new integrated amplifier, A&R was swift to respond. John Dawson 'wanted a piece of this action' and was determined to come up with a product that 'clearly caught the imagination of the customer'.

The Arcam Alpha 10 was the result. It is a powerful stereo amplifier whose modular design allows it to be expanded into a multi-room control centre. It can alternatively be transformed into a home cinema amplifier. Not only did the Alpha 10 win Millennium Product status, its versatility and performance also went down exceptionally well with the press, retailers and consumers, paying back John Dawson's determination to create a product that outperformed similarly priced rivals.

John Dawson takes responsibility for strategic leadership while detailed work is left to his ex-wife, Jacky. 'I'm not much of a committee person. We divide it up,' he explains. 'I look after the sales people and drive the design at the highest level. She does the day-to-day driving. I tend to throw in ideas to put them off course, though occasionally they're good ones. She looks after manufacturing and finance.'

The company is now owned 40 per cent by Venture Capitalists. John and Jacky own the remainder. To develop the Alpha 10, some £150,000 was spent on in-house development and £50,000 was spent on external software development. Custom case tooling cost another £50,000, and technology was also bought in from Finland for the home cinema processor.

Software development for the home cinema part of the Alpha 10 was outsourced. 'At the time we didn't have the software expertise for this complex part of the product, so we found a specialist company to provide this in the tight time scale needed. They provided us with some excellent intellectual property, which we will reuse in future designs' says John Dawson. 'However, managing such contractors is an art in itself and, in order better to control our own destiny, we have since taken the strategic decision to strengthen our in-house software team, in order to become less dependent on outside sources.'

A&R prefers to employ people with a couple of years practical experience, rather than just seeking those with appropriate qualifications. Staff work in teams and most people are paid on a salaried basis. Currently the management team of eight is salaried and paid a share of profits; this is soon to be modified with a profit sharing scheme for all employees.

Dawson believes that you have to drive what you want back down the organisation. He regards himself as an entrepreneur and the outside face of the company. It is clear that he has detailed knowledge of the market and of A&R's position within it.

He drove the concept of the Alpha 10 through the company. 'It took a lot of very able people to bring the product into reality.' Dawson formed an in-house new products team. This involving home and export sales teams, together with teams to handle industrial design, engineering and manufacturing. 'They came up with many good ideas although I would still say I am the main entrepreneur.'

Project team members were encouraged to explore new ideas in an informal, unstructured way.

case 2

licence to explore

Once JCB had decided to expand into industrial handling vehicles and had committed itself to building a forklift truck for the first time, Sir Anthony Bamford, the company's chairman, told the team to plan for the launch of the new product at a major trade show, even though the concept was only three months old. The market might have been over-supplied and immediate returns looked risky, but the forklift truck project team was driven by the chairman's commitment to develop an innovative range of material handling equipment.

'Internal promotion' was carried out among managers within JCB and its suppliers to ensure they 'bought in' to the new vision of the forklift truck at an early stage. They recognised a different approach was needed to succeed in a new market sector. Project team members were encouraged to explore new ideas in an informal, unstructured way. Provided project milestones were passed and the original schedule was adhered to, Sir Anthony Bamford was prepared to give free licence to the project team.

JCB is very open with its staff. This makes everyone directly accountable for their actions, while also ensuring the ultimate decision-makers can look beyond top-level financial data and buy in to strategic decisions such as the forklift truck initiative.

A total of 350 staff work in the Special Products division, which manufactures a range of machines, including the Teletruk, the new forklift. The workforce is young and committed, with a very low level of staff turnover. There is no union, but a works council ensures all staff enjoy identical conditions and benefit from good facilities.

Project teams normally have six or seven personnel, who select their own team leader. The multi-disciplinary team for the Teletruk project included an engineering project leader, cost estimator, industrial designer, dealer representative, representatives from the production, parts and service departments and somebody with previous experience of selling forklift trucks. The core team was gradually increased to almost 30 – though each addition had to be fully justified to the company chairman. JCB does not give project managers separate offices, thereby ensuring everyone works openly as a group.

case 3

fourth emergency service

From the start, Gary Meadows, project manager, understood the importance of employees 'buying in' to the new system.

The AA was burdened with a call-handling, recording and dispatch system that was bursting at the seams. A new technology-based system temporarily dealt with increasing demand and cut the number of call centres from 27 to seven. However, it was clear that the simple adaptation of a paper-based system to a technology-based one was not enough. The AA decided to start from scratch.

A new system was proposed that could cope with high transaction rates, maintain reliability and automate many of the business processes. Initially, it was felt to be too radical. But the project manager quickly took steps to manage risk, organising independent evaluations and presenting the Board with a rolling four-year programme that incorporated milestones at which the project could be stopped. By degrees, the Board gave its approval.

Over the first two years, the technical infrastructure was put in place. Then business processes were improved; call handlers were empowered to send specialist vehicles immediately and the deployment decision-making process was automated.

The AA's innovative approach has reduced the number of deployment centres to three. Not only has 'AA Help' won Millennium Product status from the Design Council, it has also created unprecedented levels of staff and customer satisfaction and firmly established the AA as the 'Fourth Emergency Service'.

From the start, Gary Meadows, project manager, understood the importance of employees 'buying in' to the new system. Anticipating resistance from those whose jobs would be automated, Meadows did not reveal the full extent of the project until the system was in place. Emphasis was placed on the way the system would help people do their jobs better, rather than taking away the skills they had. Once the system was in place, a culture of communication was established to help employees understand and buy into the project. Within 18 months, users were enthusiastically acting as champions for the cause.

The project was designed entirely in-house; no specialist designers were used, despite exploring products from third parties. Nigel Brown, who had been with the AA for a year and had experience in software design, provided the original design. He and Meadows were given considerable freedom to pursue the project. Call handlers and road patrols, not members, were viewed as customers, and were strongly supported. Brown and three key programmers were on hand to quickly fix any system problems, which had a positive effect on user confidence.

As the project team was not able to foresee the impact of the changes, they adopted an experimental, pilot approach by setting the system up in the Lake District, then rolling it out to other areas. The Lake District was selected because it was the personal responsibility of Gary Meadows, who would protect and support it, rather than 'blowing the whistle' if any difficulties surfaced during development. The team, which included 25 programmers, continued to 'fine-tune' the system. These improvements carry on to this day. Feedback from users who worked alongside the team was core to the way the system worked.

Meadows and Brown learned that they had to act as champions of the cause and adopted a style of thorough communication with all users. In turn, they were careful to ensure support from the Board by managing expectations and were eventually given considerable freedom to pursue each phase of the project with full Board approval. Staff reaction to the system was monitored through surveys, and meetings and staff were strongly supported.

The AA board was never presented with a very large investment decision. The initiative was always portrayed as a rolling programme. Clear business benefits were attached to each small part of the process. Rather than having to take strategic decisions on a major investment programme, senior personnel only ever had to make smaller-scale tactical decisions.

When the project was initially proposed, ICL (with whom the AA had close links) claimed the plan could not be realised. As the AA was traditionally risk-averse, Meadows did not expose many of the innovations for some time. Expectations were managed to offer only as much as would be needed to justify the project – even where Meadows believed that greater benefits were likely. 'We know where we want to be in three years' time,' says Meadows, 'we know how to get there and we will drip feed improvements along the way.'

Board approval was swung when independent experts endorsed the design; these experts were carefully chosen because they were likely to be sympathetic to the approach. Meadows also protected the testing and ensured its success by piloting it on his home patch. While any AA members adversely affected by the pilot were generously recompensed, any problems were kept quiet!

An inter-disciplinary team was set up in the IT department to implement the project. At its peak it involved 60 people; 25 were programmers and the rest were users, technical support and external experts. There was a core management team of five that kept close control over the architecture. The team set out the structure, with users designing the details. In conflicts, management had the final say. The process was controlled using the formal PRINCE project management methodology. This close linking of business and technical vision was key and stopped the project from drifting off course.

2.3 new ways of working

'only businesses that are willing to drop a number of the "old ways of working" will succeed in overcoming the challenges presented by the dramatically changing new world that we now live in'

**Nigel Crouch,
Senior Industrialist,
Innovation Group, DTI**

contributor profile

nigel crouch

**senior
industrialist**

Nigel Crouch is a Senior Industrialist, who has run businesses for the past 19 years and been committed to innovation throughout his career. He has held a number of senior marketing positions with Cadbury Schweppes, both in the UK and internationally, and was on the Board at Reckitt & Colman as New Business Director of their Household Products Division. He moved into general management with Ciba and was then Managing Director of the Evo-Stik Adhesives & Sealants business of Evode Group for almost nine years. He currently runs his own investment business and spends part of his time working with the Innovation Group of the Department of Trade & Industry.

At the DTI, Nigel is heavily involved in several major on-going leadership and innovation programmes, working closely with a comprehensive network of external partner organisations. In conjunction with the Design Council, he has played a pivotal role in both the fieldwork research and roll-out of the powerful 'Living Innovation' findings, which are being very positively received by businesses across the UK. He has also been very involved from the DTI side in the development of the '100 Best Companies to Work For' list, published annually by *The Sunday Times*, and based upon an independent evaluation of the real perceptions people have of the organisations they work for and, allied to this, helped lead 'Partnerships with People'. This is part of the Fit for the Future campaign and one of the most in-depth investigations of recent years into how a number of highly successful organisations have managed to bring the best out of their people to significantly enhance their bottom-line performance.

Copies of 'Living Innovation', Partnerships with People' and the '100 Best Companies to Work For' can be obtained free from DTI, Admail 528, London SW1W 8YT or on orderline telephone 0870 150 2500 and fax 0870 150 2333. Additional information is available at www.dti.gov.uk/pwp; www.lidiagnostic.com; www.livinginnovation.org; www.sunday-times.co.uk/100bestcompanies

2.3   new ways of working

Only businesses that are willing to drop a number of the 'old ways of working' will succeed in overcoming the challenges presented by the dramatically changing new world that we now live in.

A recent survey found that in a single day last year there was as much world trade as during the whole of 1949. In a single day last year there was as much scientific research as in the whole of 1960. In a single day last year there were as many telephone calls as during the whole of 1983; and in a single day last year there were as many e-mails as during the whole of 1990. In this highly dynamic environment, effective innovation is key not only to growth but to very survival itself.

A lot of things that have applied in the past are simply no longer valid today. Ideas dropped down from senior management in the higher echelons of organisations onto the unsuspecting workers at the coalface will *not* provide the necessary creativity essential to meet ever more stretching customer demands. Enlightened innovative companies go to great lengths to really tap into the creativity of all their people – and particularly those at the sharp end. They genuinely involve everyone in where the business is going and they actively encourage all their people to continually come up with new ideas, while making sure that proper feedback is always provided. They also manage in a lot of fun at work, which is an excellent indicator of the level of innovation within an organisation.

They recognise that good ideas do not always come when you hope and expect them to. Over and above finding ways to capture those flashes of inspiration, which strike during long-haul flights or motorway hold-ups or, closer to home, in the shower or at three in the morning, they work hard to ensure that the right people are in close proximity to each other in the workplace to spark off each other – including strategic location of the coffee machines to achieve maximum spin-off. At Creature Labs in Cambridge, the local Henry's Wine Bar was so key to the development of their highly successful new 'Creatures' computer game as a result of the team's lunchtime and after-work sessions there that Henry's was credited on the product packaging for its significant contribution to the innovation!

These ground-breaking organisations are also willing to take much more risk than is the case in more traditional outfits. 'We recognise that one product of de-risking the business is not following your hunches.' It should be stressed that they are very rigorous about evaluating the risk potential but they will then take a sensible degree of calculated risk to ensure that they do make the necessary strides forward. As one director of a plc put it: 'We felt we could and should do it but that it wouldn't be easy. We committed on the basis of half the evidence. If you don't do that, it's easy to do what you've always done.'

Allied to this 'having the guts to go for it' factor, they are also aware that, while every organisation must have in place the appropriate controls, guidelines and principles, equally there are certain occasions when intuition and experience dictate that the rules can be 'bent'. It was expressed succinctly by one corporate: 'For successful innovation you have to have an element of breaking the rules or at least bending them.' Within the Living Innovation programme there are vivid examples of the rules being 'bent' in the interests of successful innovation. Teams quietly persevere, often in their own time, to overcome serious problems on new product projects formally stopped by the Board and succeed in doing so and get their projects reinstated. 'A couple of staff were so motivated, they carried on regardless.' Senior management 'bend' their own rules on go/no go capital investment decisions as a direct result of the passionate belief of the project team that minimum acceptable financial criteria can and will be achieved and, ultimately, are. 'If the strategy is right, we will get the numbers right.'

And when success follows success as it inevitably does, as a result of these new ways of working, another old British tradition is cast aside, namely our embarrassment at anything associated with winning. These innovative companies take great delight in their people celebrating their success. Recognition of great performance is a very powerful driver and it is by no means just financial recognition, important as that is. A genuine and timely personal 'thank you' is a tremendous motivator. One Managing Director puts 12 small shiny pebbles into his right pocket every morning. He spends a significant amount of his time walking around his operation informally talking to his people and, every time he can say a genuine 'thank you' for something well done, he does so and moves one pebble from his right pocket to his left pocket. If he has not got 12 pebbles in his left pocket by close of play, he is very disappointed. At Dollond & Aitchison, peers nominate 'incredible colleagues' who, if successfully nominated, are invited to live out one of their 'incredible fantasies' – within reason! – courtesy of the company. It could be driving a racing car around Silverstone or making your first parachute jump. Whatever it is, what a wonderful way to say 'thank you', which that individual will never forget.

One other big difference between the 'old' and the 'new' is understanding that, while hard work and good planning are essential pre-requisites, successful innovation also relies on a not insignificant degree of serendipity and good luck – and being quick to ride this whenever it occurs. It is also true, of course, that great innovating businesses also make more than their own fair share of good luck!

The final message from the 'new' to the 'old' is from another respondent on the Living Innovation Programme talking about innovation: 'Don't do it unless you have a passion for it'.

1

case 1

**intimidated
no more**

Instead of simply selling products, the culture of the company is now focused on customer satisfaction.

Innovation has transformed performance and profitability at the optician Dollond & Aitchison. Its corporate culture now encourages a continuous stream of creative ideas and the operational structure has been transformed to permit rapid implementation of innovation.

It was not always like that. When Dollond & Aitchison first decided to find out more about customer needs, it employed specialists to conduct ten focus groups. It learned that people were overwhelmed by the choice of glasses, intimidated by eye tests and threatened by the black leather examination chairs. Buyers knew they had to select glasses that would suit their faces, but had no idea about the structured processes involved and therefore had to rely on trial and error.

On a branch visit, chief executive Hardy was lucky to meet a manager who had noticed more and more of her customers were employing independent image consultants to help them choose their frames. In turn, she had successfully run a series of evening sessions with an image consultant to test this service in branch on her customers. Hardy recognised the business potential of the concept, and the store manager was transferred to head office to work alongside a leading professional image consultant to develop the idea. Working in secret, they jointly developed customer profile forms, colour guidance materials and visual aids to help staff determine the shape of each customer's face. Positive feedback from an in-store trial led to a decision to invest £4000 in a Styleyes system at each sales outlet, demonstrating to everyone that the old rules of the company were being completely rewritten.

Fundamental changes were then made to Dollond & Aitchison's retail premises. To make them more customer friendly, they were painted a soft mint green. The threatening black chairs were replaced with green versions and the dispensing areas were separated for privacy. Customers are mailed a feedback form to evaluate every stage of the buying process. Despite the lack of any incentive, 16 per cent of customers respond. Each form is logged and directly linked to each branch. Instead of simply selling products, the culture of the company is now focused on customer satisfaction.

Suppliers also visit the Dollond & Aitchison head office at least once a week to get feedback from the Styleyes programme about gaps in the market. The company employs a frame development team to encourage suppliers to respond to customer demands and needs for product innovation.

case 2

creative friction

A flash of genius at home led not just to a breakthrough in friction stir welding. It also reinforced TWI's efforts to encourage more freethinking among its 400 technologists.

TWI aspires to be a world leader in welding and joining technology. From a site outside Cambridge, it produces a stream of sophisticated refinements to existing techniques and processes.

A flash of genius at home led not just to a breakthrough in friction stir welding. It also reinforced TWI's efforts to encourage more freethinking among its 400 technologists. The danger for a large research organisation like TWI, argues its principal research engineer Wayne Thomas, is that technologists are so busy producing modifications and adaptations they never take a step back. To put it another way, they never see the wood for the trees. He finds it hard enough himself. 'It is difficult to innovate here. The phone's ringing all the time or there's some interruption.' Not surprisingly, his moment of inspiration happened at home in the shower.

Wayne had been working on friction welding for 23 years, and has been responsible for the conception and development of a number of emergent technologies. What dawned on him was that an important step in the process had been overlooked.

Traditionally, friction welding involves two bars pressed together in a machine where one is rotated under pressure. Friction stir welding is a radically new technique. A high-temperature rotating tool moves steadily along the joint line forming a plasticised zone that coalesces behind the tool to form a solid-phase weld.

TWI has a budget for blue-sky research, allowing ideas from both employees and member companies to be researched and exploited, although it recognises its shortcomings in fostering ideas. 'Brainstorms are useful,' says Keith Johnson, Associate Director, Contract R&D. 'You can get incremental solutions to problems but, if you want a quantum leap, it generally comes from one person. The flash is unlikely to come to two people at the same time.'

Creativity sessions have been set up to encourage these flashes of inspiration and two researchers have been taken off day-to-day project work to drive forward the innovation process in TWI. 'There are a lot of highly specialised people here', says Wayne Thomas, 'but the perennial challenge is being able to see the wood from the trees. You have to be highly specialised to come up with an innovation. You also have to have a broad based background as well, or you can't latch on to these things.'

TWI is a company limited by guarantee with a turnover of more than £20m. It is supported by company membership fees and research project funding. It has 2500 corporate member companies world-wide and 4400 individual ones. In 1998 its EU-funded collaborative projects were €4 million value.

Incentivising innovation is a bit of an issue from the employee side. Although it is reflected in performance increases, rewards are not as direct as in Germany, for instance. Recognition occurs within the company and some nominal ex-gratia payments are given. Wayne Thomas has also received a national award for innovation.

'We are all creative. That's the point that we should get across. We do it every day,' says Wayne Thomas. 'Lots of people can do that. What we are trying to focus on is the step changes, giving high value invention.' In order to achieve this, work at TWI is accomplished in teams that change according to the project. No re-training was required as a result of work on friction stir welding.

'There is a gap between the present day and the future. If you ask people to look at that gap, they talk about ten years' time. What I think of is a nanosecond away. We try to capture creative thought. You've got to go back a stage and look at creative thinking that leads on to things like friction stir welding. That is an approach that we are trying to pass on to other people.'

2.4

risk management: delivering a project on time, on budget and defect free

'the way we traditionally manage projects and risks is fundamentally flawed'

**Adrian Blumenthal,
Development Director,
Rethinking Construction**

contributor profile

adrian blumenthal

development director

Adrian works for Rethinking Construction on secondment from Crown House Engineering, one of the UK's largest building services companies. Rethinking Construction is the UK Government sponsored construction industry improvement programme (details can be found on www.m4i.org.uk). Rethinking Construction's programme spans all sectors and markets of the construction and manufacturing industry. Adrian works alongside many of the 300 (£6bn) construction industry demonstration projects in London, Northern Ireland, Scotland and the South East — delivering the business case for better industrial performance in line with the targets laid out in the report Rethinking Construction. Adrian has been involved with the strategic development of Rethinking Construction, bringing many parts of a fragmented industry together to share, learn and evaluate best practice and innovation.

Adrian considers himself as a practitioner and works alongside those who want to learn how to improve their performance. He is also is an accomplished innovator in his own right, delivering some of the most productive sites in the world along with some new product innovations.

risk management: delivering a project on time, on budget and defect free

Project management today is a very complex affair with many contracts still procured on the lowest bid and the baggage that may come with it. Worse still perhaps is the move to long-term partnering contracts, or framework agreements that derive from a solutions-based approach – basically 'you are with me because of the problem', which on the face of it is a panacea, but the collateral of a failing risk in this type of relationship can be much worse.

Because of the 99.9 per cent likelihood that the project will fail, this has brought up the question of the competence of those involved to manage risk in projects effectively. One wonders what it must be like for those who have to work in both environments simultaneously. What that means in practice is that the project team is both the arsonist and the fire fighter; maybe this is why industry at large is schizophrenic.

Despite the way a contract is brought to fruition, as mentioned above, many projects still share the same elements of risk and we all know that the projects themselves are the vehicle for the risks being discharged. The way we traditionally manage projects and risks is fundamentally flawed. Construction project and risk management is a reactive process, which is carried out with the maximum economy of effort – you just have to visit any site in the UK to see this in action.

It is most likely that as companies embark on implementing *Rethinking Construction*[1] and the principles of the report *Accelerating Change*, the current methods and science of delivering projects may be deemed inappropriate. The collateral for not methodically evaluating risks is relative and maybe quite substantial – just look at the increasing trend of accidents, transaction costs and litigation the construction industry faces each year. What is really disturbing after four years of *Rethinking Construction* is that most performance measures still concentrate on task completion rather than effective process management. I do not know of any client or team that that would not want a project delivered earlier, with zero defects and accident free. Project management is about managing uncertainty and therefore risk.

There are many types of risk and perhaps the most important other than safety (particularly in the construction industry) is not delighting the customer and client in the delivery of their project. The goal of any project is about satisfying, or enhancing if possible, the client's value proposition that mitigates their own risk and delivers their ongoing business security (something many of our colleagues seem to forget). Typically through current project delivery techniques (most based on the concept of *real politik*) some of the customer's productive requirements are often missed, deemed too expensive or simply overlooked. Perhaps there is a way of managing contracts to over come this perennial problem.

Rethinking Construction challenged the industry to think differently – in terms of organisation, people and processes. It challenged the industry to adopt the logic of Right First Time as a driver for change. It dictates that if SOPs (Standard Operating Procedures) are in place and FMEA (Failure Mode Effect Analysis – Operation and Business Process) has been used in their creation, there can be no defects and therefore no accidents can happen. Easy in theory, quite the opposite in practice I'm afraid, as there are still a huge number of sacred cows out there.

The root cause of the problem is the way we manage uncertainty (or risk) on our projects – or perhaps the way we mismanage uncertainty on our projects. Uncertainty comes in many guises because of the way we measure our performance on projects and to cover ourselves we tend to add huge amounts of safety in projects task duration. We tend to plan with up to 50 per cent added safety in task duration and couple this with variation in task duration that can vary between minus 80 per cent to plus 500 per cent for a 10-day activity. It begs the question how do we effectively manage risk in projects. With this amount of waste and the fact that value-added is typically 0.5 per cent for each activity on the construction plan, then there is scope to improve materials and information flow during each stage of the construction process. This allows us to plan better, reduce risks created through variability and also allows more time to design out risks and finish on site earlier.

---

[1] *Rethinking Construction*, Report by the Construction Task Force, 1998.

If you do not understand how value is created in process time, then you can not reduce the variation in the process and construction plan. Therefore you have lost control of the project, and risks increase as do the chance of accidents. The industry only has itself to blame. It is the industry that creates the contracts and oversees the downfall of the projects. Who has ever worked on a contract that was delivered embarrassingly early, defect free, risk free with zero accidents and, at the same time, made a killing on the margins and delighted a client? If you are out there, then let us know.

Initial projections for JCB's attempt to break into forklift trucks suggested that its Teletruk would not be profitable for some time.

The company was prepared to invest several million pounds in the belief that cost difficulties could be overcome during its development.

case 1

tackling cost difficulties

JCB's confidence partly sprang from its practice of using as many existing components as possible. For example, the single telescopic arm that makes the Teletruk so different from rival forklift trucks, is fundamentally the same item used on other JCB loading machines. Many of the components are shared with other JCB vehicles including the engine and transmission. This brings obvious benefits to the efficiency of the supply chain and it reduced the risks involved in producing new components.

Local companies, partner organisations and suppliers all helped to evaluate the forklift truck during the development process. In order to ensure competitors did not learn about the novel design features of the Teletruk, the evaluators were all obliged to sign non-disclosure agreements. The company began to build contacts in the forklift truck market by mocking up promotional material showing a 'Trojan Horse' JCB forklift truck that was entirely conventional and gave no clues about the novel design features of the Teletruk.

Numerous JCB staff from other departments climbed in and out of an adjustable prototype, sat in the seat for long periods and used the controls to provide feedback on areas where improvement was necessary. JCB wanted to ensure forklift truck users would recognise the Teletruk was in a class of its own.

There was a formalised project plan controlled using Microsoft Project tools, in which over 700 activities were scheduled. The project team had to report to the executive board at various stages to prove that key milestones had been passed.

Developing a new product for a new market sector also demanded a high level of flexibility. Production levels had to be adjusted every week to meet varying demand patterns and short delivery promises. Temporary staff were used as a buffer whenever required.

Moving into a new market sector also presented sales challenges. Existing JCB dealers were experts in the construction and agricultural sectors. So JCB had to set up a telesales operation to generate awareness and interest in this unfamiliar market sector. It is also appointing new dealers, and has created new financial products to suit customers in the sector who prefer to hire or lease, rather than purchase, forklift trucks.

From the very beginning of the Teletruk project, JCB carried out market research among forklift truck users, customers, suppliers and even a specialist forklift truck training company. Comparisons with existing forklift trucks focused on the five top-selling models to evaluate everything the market leaders could do. To ensure JCB knew as much as possible about the new market it was going to enter, it joined the forklift truck trade association and the Institute of Logistics. It also brought in rival products for evaluation and visited future rivals' factories.

As the Teletruk project developed, a design consultancy advised on look and feel. The objective was to ensure the Teletruk's ergonomics would satisfy 95 per cent of user sizes and shapes. All members of the Teletruk project team learned how to drive forklift trucks and worked around their own factory to gain first-hand experience of the difficulties faced by users.

The company learnt the hard way not to promise before it could deliver.
BioProgress Technology was set up to find a cleaner way of disposing of nappies. It used biodegradable materials to make nappies that could be flushed into the sewage system, instead of clogging up landfill sites. Although the idea was well conceived, the nappies encountered difficulties in breaking into the UK market. BioProgress's efforts did not go to waste. It has absorbed its lessons and built a successful business using biodegradable materials for a variety of sanitary products.

A combination of difficulties in production and distribution scuppered the attempts of BioProgress to break into the UK nappy market. In moving from idea to production, BioProgess found that its raw material costs were 15 per cent more than traditional nappies. More problems than anticipated were encountered in contracting out production and there was fierce competition from market incumbents such as Procter & Gamble.

Various multiple retailers proved very difficult to penetrate, due to price and cost barriers, so the company then focused on mail order and small pharmacies. Mail order was a minor success, but the product's bulk compared with the sales price meant that retail outlets could sell many more smaller and more expensive products such as cosmetics for the same shelf space.

The company also learnt the hard way not to promise before it could deliver. Once a prototype of its biodegradable nappy had been developed, BioProgress agreed to attend an awards ceremony. It felt that a demonstration of the product would help a bid for an additional grant of £48,000. The successful demonstration attracted real media interest and before long BioProgress was performing demonstrations on a number of TV shows, including *Tomorrow's World* and *The Richard & Judy Show*. Overnight the switchboards were jammed with hundreds of calls from mothers – 12,500 in five days – all looking to get hold of the revolutionary new nappies. A great deal of time was spent explaining that they were yet to go into production.

After not succeeding with the nappies, BioProgress looked at other applications for its technology. On the basis that consumers were prepared to pay a lot more for sanitary towels than for nappies, the company saw this as a more lucrative market and developed a biodegradable sanitary towel, Harmonies, which could be flushed away or thrown onto a compost heap.

In selling Harmonies sanitary towels under licence, the company's knowledge of protecting intellectual property rights has been paramount. 'Not manufacturing our principal asset,' says Barry Muncaster, the company's chairman, 'means that our intellectual property has to be protected at all costs.'

Barry Muncaster is frustrated by the need to spend so much time on bureaucracy that neither protects innovation nor assists its commercial exploitation. Small companies, such as BioProgress, face a number of problems before they get down to the business of innovating, he argues. Bureaucracy is the biggest burden, says Muncaster, observing that yearly returns to Companies House cost £32 a time only to say that nothing had changed.

BioProgress was approached by a Canadian company that was prepared to produce and sell Harmonies sanitary towels under licence in Canada, which is a very environmentally friendly nation. Harmonies was subsequently the only product in its class to be awarded the Canadian government's EcoLogo award. Success in Canada led to the sale to the Canadians last year of the US marketing rights for $5m plus royalties. Applying the same technology, the venture is now expanding into the panty-liner market and revisiting the nappy concept through the distribution opportunities of the Internet.

2.5 intellectual property

'knowing how each law works, and what it protects, is the key to making sure your creativity appears on your balance sheet as an asset'

Anthony Murphy,
Director of Copyright,
The Patent Office

contributor profile

anthony murphy

director of copyright

Anthony has been a member of the Senior Civil Service and Director of Copyright in the Patent Office since September 1999, and since May 2002 he has also been responsible for the Patent Office's marketing and information activities. He is a member of the Patent Office Board and a member of the Council of the Intellectual Property Institute. He is also on the Editorial Board of *Intellectual Property Quarterly* and the British Library's Advisory Committee on Science, Technology and Business, and chairs the Government's Counterfeiting and Piracy Forum.

Anthony Murphy joined the Civil Service in 1992 after 16 years in the private sector, spent mainly in Asia and the Middle East. During this time he was an adviser on currency management issues to the Association of South East Asian Central Banks and, as an Associate Director of the De La Rue Company plc, he was responsible for developing new markets in Eastern Europe and the Former Soviet Union.

Anthony was co-author of the Government's 1993 *Review of the Implementation of EU Law in the UK,* before being seconded to the Foreign and Commonwealth Office as First Secretary (Industry and Competition) at the UK Permanent Representation in Brussels. After three years he returned to the DTI to become Head of Asia-Pacific Trade Policy, representing the UK on the Asia-Europe Investment Experts Group and other trade-related bodies.

He read Modern History at Oxford University, and is a Freeman and Liveryman of the City of London. He was recently made a Fellow of the Royal Geographical Society in recognition of his work in Africa and Polynesia.

intellectual property

Innovation can take many forms, each protected by different laws, known collectively as intellectual property rights. Some of the rights afforded by IP are free and automatic, and some have to be registered. Knowing how each law works, and what it protects, is the key to making sure your creativity appears on your balance sheet as an asset.

Most obviously any business needs to protect its good name – today's biggest brands have the majority of their value in their trade marks. Any trader who finds that a competitor is using the same or similar name in trade, so as to confuse or divert customers, can bring a so-called 'passing-off' action under common law. However, Registered Trade Marks (RTMs) give much more powerful protection, including both civil and criminal remedies, and are available for people who register their trade marks at the Patent Office. A registered trade mark, renewed for £200 every ten years, can last for ever, and gives the owner the right to stop others who trade in the same goods under the same or similar name, whether or not they do so in ignorance of the registered trade mark. A good registered trade mark is their 'badge of origin' – distinguishing them in the market from their rivals.

It is crucial to check, before you start trading, that the name for your business has not already been registered by another. The Trade Marks Register can be searched for free from the Patent Office website. Once you are sure that your trading name does not infringe an earlier registered trade mark, you should consider registering your own trade mark before anyone else does.

Once your business gets started, you will quickly create a body of knowledge unique to, and critical for, your enterprise. These are so-called 'trade secrets', and they are protected by the law of confidence. You cannot formally register a trade secret, so it is up to you to protect information that is vital for your competitive edge. Staff, contractors, investors, suppliers or anyone else with possible access to the business' secrets should have their relationship with your business covered by a confidentiality agreement. This will give you the right to sue if anyone leaks details of secret customer lists, business plans, marketing strategies, pricing schemes etc. Staff who have business-critical information in their heads can be prevented, by a so-called 'restrictive covenant', from working for the competition within a reasonable period from leaving the business.

For those whose business is either manufacturing or research and development, patents and registered designs are of crucial importance. Patents and registered designs can protect a product, either in how it functions, or how it looks. These rights have a limited duration, but like a registered trade mark, allow you to take action against copies even if the infringer did so in ignorance.

If you make a product which is ordinary in its functionality, but styled in an innovative way, for example a kettle with an elegant spout and funky handle, then it is likely to be a market success not because of its function (which is routine) but because of its looks: this is what design registration protects. An application for a design registration will only succeed if the design is novel (new) and has 'individual character', in other words it must not be substantially similar to existing designs on the market. A designer seeking registration must apply within one year of publicly disclosing his/her design, otherwise his/her showing of the design to the public counts against his/her application for lack of novelty. The Designs Registry is at the Patent Office, and the first five years' protection cost £60, renewable up to a maximum of 25 years.

Patents protect the functional technology behind innovative products or processes, but think carefully before applying for patent protection. Clearly, if you put a product on the market and it is a success, competitors will buy it and copy it, so in this situation a patent would be a good idea. But if your innovation is a novel process, and the final product on the market gives no clue as to how it was made, then that process might be better protected with a trade secret. You see, part of the patent 'bargain' between the inventor and the State requires the invention to be disclosed completely to the Patent Office, who publish those details 18 months after application for everyone to read. This makes a patent the exact opposite of a trade secret. The reward for that disclosure is the right, for up to 20 years, to exclude others from making, using, selling or importing your technology unless they buy a licence from you (or you sell the patent).

All technologies have patents on them, from machinery to medicines, electronics to engineering. Patents have protected the most important inventions of our age, the jet engine, the hovercraft, the internal combustion engine, but also some of the most modest: the lid to a jar of peanut butter is protected by a patent because of the way its underside has been shaped. But to have a patent granted the invention must be novel and inventive. This means that it must not have been publicly disclosed to anyone prior to the filing date of the patent application. If the Patent Office finds that your invention has been published before in an earlier patent or journal, or if you have shown the invention to others without a confidentiality agreement in place, then your application will fail. To get a patent the Patent Office needs to be sure that the alleged 'invention' is not merely a combination of obvious features or steps: they want to see an inventive step.

The patent application procedure involves many hurdles, which take several years to clear. The initial application can be filed for free, and this secures a so-called 'priority date' after which it is safe to disclose the invention because only disclosures made before the priority date can attack the novelty of the application. At the end of the first year you need to decide which other countries you want protection in, and file applications in those other countries, who will treat those foreign filings as if they were received abroad on the UK priority date. Meanwhile, to advance the application at home, £130 needs to be paid to have the application 'searched' and published. This means that the Patent Office looks through databases of earlier patents and journals to see if the invention has been done before. The results of the search will be sent to you, and your application published and added to the databases for everyone else to look at. A year later, if you pay a further £70 for your application to be 'examined', its legal and technical detail is closely scrutinised; only claims to new and inventive technology will be allowed. If you request both search and examination together (cost £200), this will speed up the whole process. After a few exchanges of correspondence to amend the application, it will be granted.

Although the total Patent Office fees up to grant are merely £200, there are other costs. Once a patent is granted it must be renewed every year to keep it in force, otherwise it lapses and becomes free technology for anyone to use. In the UK, patent renewal fees start at £50 for the fifth year, ramping up to £400 for the 20th (final) year. Then there are patent agents' fees. Patent agents help with the legal detail of the patent application procedure, in much the same way as a solicitor deals with the legal side of buying or selling a house. Their fees can run to several thousand pounds. And if you want patent protection abroad you need to pay for agents in those other countries, the official fees of the foreign Patent Offices too, and pay for your application to be translated into the other languages. Current average figures for patent protection around the world estimate it takes £6,500 to get a US patent, £10,400 for a Japanese patent and over £31,700 across Europe once all the translation costs are factored in.

Anyone struggling to fund patent applications in a handful of markets has to wonder whether they could afford to defend the patent in Court were it to be infringed. Even simple disputes can cost over £200,000 if not resolved out of Court. An inventor with modest means may find that the best course of action is to sell the patent application at an early stage to a big company with deep pockets, or to licence the technology to a big manufacturer in exchange for a modest royalty and an agreement that enforcement costs will be met by the licensee. An alternative is to buy litigation insurance – just having the insurance can make would-be infringers shy away.

Even if your company is small or just starting out, merely lodging a patent application can be a good idea. Many venture capitalists won't even consider investing in a new technology until they know that you have some patent applications in the pipeline. The application itself will have a value on the balance sheet if you need to sell your business.

Ron Hickman started out on his own with his patent for the 'Workmate®', but licensed the patent to Black & Decker for £1 per unit – and they sold 30 million worldwide whilst the patents were in force. A small company with a patented technology vital to bigger players can find itself in the happy position of being able to 'punch above its weight', entering into lucrative licensing deals with those larger firms who have substantial markets that the small company could never hope to reach on its own. The trick is to cope with potential infringement, and that is more to do with good business sense than the patent itself. When it comes to being on the receiving end of an allegation of patent infringement, for something you have been doing for years, you will be grateful you had the wisdom to at least have had your technology published in a patent application (even if you never took it any further) because that disclosure will invalidate any later patent for the same technology sought by another. This is a 'spoiler' tactic – you disclose your technology without pursuing patent protection for yourself just to be sure that no-one else can have a patent for it either.

There is no disguising the expense of the patent system but there is a flip-side to the patent coin which is free. Because every patent application contains a complete description of someone's technology, and because patent applications are published, and now appear in on-line databases, you can trawl these patent applications for information vital to your own research and development efforts. Why struggle to solve a technical problem already solved by another and published in an application? It is estimated that 30 per cent of European R&D is wasted re-inventing technology already disclosed in patents.

The 'Esp@cenet' database at the Patent Office website is free to search and has over 30 million patent applications in it. If you find a document that describes your would-be invention don't be disheartened. This earlier document merely stops you having patent protection for your own invention, but it doesn't stop you using your own invention unless the earlier patent in question is actually in force in the UK, and most are not. Many of the patent applications in the databases describe technology which is free to use, either because they were never granted, or were granted but have since lapsed because their owner stopped paying annual renewal fees. You can also use the databases for commercial intelligence to study the patent applications of your closest competitors, to see what they are inventing today to put on the market tomorrow.

Patents, registered designs and registered trade marks are all 'registrable' rights, which involve the completion of forms, the paying of fees, and the publication of your rights in a register. These are all quite different from the protection you get from copyright which is free and arises the moment a new work is created, whether it is musical, dramatic, literary or artistic. Copyright lasts for a very long time too, typically for the lifetime of the author plus 70 years.

Copyright is not an alternative to patents etc, and the protection it affords is not the same as that for the registrable rights. For a start, copyright does not protect a short phrase like a company name, because a mere name or title is not in itself a 'literary work' like a poem, script or novel. And although copyright may protect blueprints, technical diagrams and instruction manuals for an innovative product or process, that protection only gives a right in law against people who, for example, photocopy such paperwork: the technical principles embodied in the invention itself, and as protected through a patent, are not covered by copyright. Furthermore, although a one-off work of art may be protected by copyright, once an article goes into mass-production copyright ceases to apply, and protection should be sought through design registration. Unregistered design right, which is free and automatic, arises when a new design is created.

At least copyright is free from any registration hassles and paperwork. Copyright arises very easily, but its existence and its ownership become difficult to prove when allegations of copying start to fly, because there is no central register to check who owns what.

Crucially, as its name implies, copyright is merely a right against copying or adaptation. If you accuse another of copying your work, for example a photograph, you must prove that they did not take their own photograph independently of you, otherwise you have no case.

With things like music or images, it is fairly simple to determine if one work has borrowed elements from an earlier one. The challenge lies in showing who did what first. You could put a copy of your work (eg: disc of software, CD of music, photo of painting, text of script) in a sealed envelope and sent it to yourself by registered post (thereby date-stamping the work), storing it in a safe place to be opened at a later date in Court if you ever needed to prove that your work came first.

You can also 'booby-trap' your work with tell-tale fingerprints which are not apparent to the casual observer, but which can be discovered in copies. If 10,000 lines of computer code have had a few lines of dummy code sprinkled amongst them, lines which don't affect the programme, and those same dummy lines appear in a rival's software product then they have some explaining to do! Recently the AA were caught out by the Ordnance Survey for copying OS maps for use in AA Road Atlases. The OS had made subtle changes to roads and streams in their own maps, changes which appeared in the AA atlases, proving that they had not used their own surveyors but had merely copied the OS maps. Damages were reckoned at £20 million!

Databases, mailing lists, technical data – all of these can be booby-trapped so as to support a later claim for breach of a trade secret, as well as of copyright.

'Patenting is usually seen as an afterthought – it needs to be upfront.'

Not many businesses in Southport have supply contracts with the US Army. But keeping American squaddies warm is just one of the growing number of applications of Gorix E-CT, invented by Robert Rix.

Patented in 1994, Gorix is a fabric that acts as a resistance-heating element that is capable of regulating its temperature without the aid of a thermostat. The even and uniform way temperature is distributed across the surface of the cloth and its inherent flexibility are two of its key features.

One of the original uses that Gorix was put to was as a therapeutic heated blanket for horses. With the ability to function from low voltages, the flexible heating element meant that there was no risk of burns or electrocution, even if the horse chewed through the cloth.

A relationship with BTG (British Technology Group) helped put in place a robust intellectual property framework, which enabled Gorix to secure major licensing and evaluation deals. Today 70 per cent of Gorix's business is with the US, including an evaluation for heated car seats for the Lear Corporation and heated steering wheels for GM.

However, as BTG moved towards flotation its focus seemed to shift from smaller operators and inventors towards larger corporate entities. The companies parted ways in an amicable fashion, and all rights reverted to Gorix. The company is now actively engaged in marketing its own licences for the technology around the world.

Rix has been the driving force behind Gorix throughout, showing 'determination and bloody-mindedness' at every stage. Having taken out his first patent for a collapsible lamppost at the age of 18 he has learnt not to take 'no' for an answer. He believes that Gorix fabric has the potential to become a global product. 'There is nothing to match it. If we get the financing right, we will be onto a big winner.'

Experience has demonstrated how vital it is to invest time and effort into the finer detail of commercial and legal agreements. Without patenting Gorix and laying down the precise terms on which other companies can use it, Rix is convinced that bigger companies would have been able to exploit his innovation with very little return for his effort.

Gorix has taken a conscious decision to limit the exploitation rights it is prepared to grant to large companies. With the aid of a law firm in Blackburn, Lancashire that specialises in textile intellectual property rights, agreements are carefully drafted, limited by geographical area and product category. Rix feels that small, high technology companies 'give away too much' to large companies by not defining product categories tightly enough when licensing. 'Patenting is usually seen as an afterthought – it needs to be upfront,' he adds.

'Customers usually come to us via our website; phonecalls, faxes and emails come in, making it difficult to decide where to focus.' No one is ruled out on first enquiry, and Gorix is prepared both to license and to enter into joint ventures. 'If companies want to have a look [at the technology], they pay to have a look.' American companies are more prepared than those in the UK or elsewhere to invest in exploiting the potential of the innovation.

There are currently two joint ventures, and active discussions are going on with other parties. First links tend to be at design-engineer level. Rix invests a lot of his time and effort into working closely with these individuals. This strengthens commercial links and simultaneously helps him to build a more detailed understanding of their technical problems and requirements.

The company needs more development capital, primarily for salaries, travel and business development, but also to be able to put on a good professional front in order to deal with huge multinationals. Relationships with bank managers have proved difficult, since they are often not proactive and find it difficult to understand Gorix as an enabling technology. Applications for SMART awards have helped to improve the focus of Gorix's business plans and technology. But lack of resources makes it difficult to gather feedback on why project bids sometimes fail, thereby hindering further business development.

Rix is the only full-time employee of Gorix, which is based in a small, inexpensive workshop in Southport. The other members of the team are all part-time. Their involvement is 'mentally very significant, but physically intermittent,' says Rix. 'The experience of my fellow directors has been essential in bringing Gorix as far as it has come. What we need to do now is to take the next big step.'

case 2

profitable
niches

'Protect your products and back this up by defending them.'

David McMurtry of Renishaw is emphatic in his determination not to get drawn into a situation where direct competition drives prices down. He insists on a corporate focus on niche markets in inspection technology where competition is minimal and margins are high.

Renishaw, the FTSE 350 company he chairs, is characterised by an aggressive commitment to innovation. The company ploughs 12 per cent of profits into R&D, it closely follows trends to find high-cost niche markets and it is exceptionally committed to protecting intellectual property rights (IPR); McMurtry is ready to take infringers to court at any time.

A lot of resource is earmarked for the protection of IPR. Patents and a commitment to defending them are introduced at the start of projects – not at the end: 'Protect your products and back this up by defending them.'

McMurtry's commitment to IPR is underscored by his insistence on keeping finances available specifically for the eventuality of a court case against a multina-tional company infringing its patents – much to the bemusement of shareholders. These expenses, believes McMurtry, are well worth paying in order to put in place a strong deterrent against any future infringements.

All stages of project development are well documented for future reference, although feedback did suggest that projects can be 'documented to death' and a balance must be struck. Realising that multi-skilled teams would be the ideal framework for product development, Renishaw has also overhauled its core structure, forming Products Divisions for each of its core market sectors.

The product range is driven by customer demand, as well as by ideas from within the company. In a sense, Renishaw is a victim of its own success; customer demand for older products is high, meaning that resources and machinery are tied up in this rather than being available for new developments.

The company range extends to approximately 200 products. Renishaw considers itself better than anyone else in these niche markets. 'Find a high-cost, fairly focused niche market,' advises McMurtry, 'and concentrate on satisfying it with the highest standards of quality and reliability.'

2.6 funding innovation

*'frustratingly for the innovator,
exciting new projects tend to
throw up unique issues for any
potential funder'*

**Michael Riding,
Managing Director,
Lloyds TSB Corporate**

contributor profile

michael riding

managing director

Michael Riding is managing director of Lloyds TSB Corporate, a business that provides a broad range of banking, advisory and financial services to the corporate marketplace. Michael joined Lloyds Bank in 1983 from Chemical Bank, New York. Senior roles within the retail side of the bank preceded his appointment as Head of UK Commercial banking in 1991. He assumed his current responsibilities in January 2000.

Part of the Lloyds TSB Group, the Corporate division delivers relationship banking expertise to companies with turnovers in excess of £2 million. In addition, and in support of corporate businesses, corporate banking is able to deliver a number of specialist financial solutions. These range from straightforward cashflow support via invoice discounting to businesses undergoing change and requiring substantial financing. Specialists from the Acquisition Finance, Development Capital, Capital Markets, Structured and Commercial Finance teams are able to understand these requirements and situations.

Lloyds TSB Corporate is based in London and Bristol with dedicated offices throughout England, Scotland and Wales, as well as a substantial presence in New York. On an international front, many of Lloyds TSB Corporate's links are worldwide, and there are extensive links with foreign banks, businesses and governments. All this enables Corporate to support business activities on a global basis. Its aim is to ensure that it matches its expertise to the requirements of the market in order for all its customers to meet their objectives.

funding innovation

Finding the funding to give life to nebulous concepts like innovation and imagination can often pose a hefty challenge for rapidly expanding businesses. The idea has been formulated, you're eager to press ahead before another party beats you to it and you think the only barrier between you and a roaring success is the cash. Think again.

Frustratingly for the innovator, exciting new projects tend to throw up unique issues for any potential funder. Research and development is phenomenally cash hungry, so income streams must be firmly pinned down. Therefore, it is prudent to take certain steps before you even consider what type of funding might suit you.

In every sphere of life first impressions are vital – and getting funding is no exception. So, the first of many hurdles to jump *en route* to bringing your groundbreaking notion to fruition is to commit an easy-to-read, logical explanation of your idea to paper. Otherwise known as the business plan, this document should pull together the common objectives for the project, details of the launch market and a run down of the management team. The part played by a company's staff and culture should not be underestimated.

Because the business plan must also contain financial information – if possible historical, forecasts and assumptions – as well as a proposed structure of funding requirements, now is the time to assess which type of funding to go for. Debt or equity, grants or venture capital? Do you have the right mix of short, medium and long-term liabilities? Is the combination of equity and debt correct? Are you concentrating too heavily on the more traditional forms of funding whilst overlooking other options?

Although debt funding – generally obtained from a bank – can be the quickest route to take if some form of tangible security is available (the Peabody Trust has leveraged its stable rental income), remember that it is possible to combine several methods and approaches. Interestingly, all three case studies have demonstrated the logic and value of mixing and matching by choosing not to rely on one source of funding.

Marks & Spencer's Imagemaster is a textbook case of a mutually beneficial relationship between innovators, higher education and the corporate world. Developing productive partnerships throughout the supply chain is truly a win-win strategy – raising standards and efficiency while ensuring contributors get a cut of any profits.

In many ways, BioProgress and Marks & Spencer have adopted similar formulae. Both have found a charismatic entrepreneurial figure to play a leading role and have found substantial grant money.

The common threads linking BioProgress' Barry Muncaster and Marks & Spencer's Charles White are: commitment, financial muscle and heavy-weight experience. The two men, who could be described as Business Angels, are well placed to contribute significantly on a practical level by taking senior positions on the project team. Unlike venture capitalists, who often insist on a presence on the board to retain control of their investment yet only offer input for a fixed period, White and Muncaster are permanent call-ups to the squad. This enables them to take the reins where necessary – without resistance from other players – to help drive work to its desired conclusion. These robust, hands-on tactics can really pay dividends.

Grants are widely available both in the UK and on the Continent. This is a fact that often comes as a surprise to many entrepreneurs and businesses – largely due to the difficulties experienced in tracking them down. Once found, grants also have the added appeal of coming unencumbered by potentially unwanted, interfering investors. Your local Business Link office is a good place to start the search, with various types of matched funding also on offer.

Whilst, historically, funding has been linked to products, there is an increasing appetite to support service sector initiatives. This could, in part, be due to the role played by new technology in the business environment of today. Ideas and projects that have the scope to develop the local business community and increase employment also tend to be more positively received.

The Peabody Trust raises the pertinent issue of initial outlay to raise funds – and the relationship between the amount spent and accrued benefits over time. Broadly speaking, grants are low cost and spend is usually complete when your application

3 Connect

3.1 partnering with customers

'do organisations produce offers that customers want or do they produce offers that they think customers want?'

William Mellis, Vice President, Cap Gemini Ernst & Young

contributor profile

william mellis

**vice
president**

William Mellis is a Vice President in Cap Gemini Ernst & Young's Customer
Relationship Management practice.

With over 20 years experience advising board level clients on major change
programmes across a variety of industries. William is well placed to identify the
benefits from the effective use of technology for major corporates.

William holds a BA (Hons) in Operational Research and Marketing from the
University of Strathclyde. Prior to joining Cap Gemini Ernst & Young, William
worked on an international basis for companies in the FMCG market place implementing business change projects using technology as an enabler.

partnering with customers

The objective of marketing is to 'sell products that don't come back to customers that do'. Taking this further, the idea that businesses design, develop and market offers[1] that customers do not want is not feasible. However, do organisations produce offers that customers really want or do they produce offers that *they think* customers want?

Most organisations have recognised this issue and as a result seek to include their customers to some extent in the development of its offers. The degree of customer inclusion can vary:

- *Consult*
  The lowest level of customer inclusion where the organisation ask its customer to respond to specific questions;

- *Discuss*
  Customers are invited to various forums to discuss predefined topics;

- *Collaborate*
  Customers are integrated with parts of the organisation for specific aspects of the offer; and

---

[1] Product and service bundles

- *Involve*
  The highest form of customer inclusion where customers are integrated with the entire organisation at all stages and aspects of the offer.

The latter is where there is true partnership between the organisation and its customers. The objective of the partnership is to maximise the value to all parties through the provision of products and services.

Most organisations are talking about moving towards collaborating with their customers, with very few, if any, organisations achieving true partner status with their customers. The good news is that there are activities that organisations can undertake to collaborate more with their customers:

- *Involve your customers in what you do.*
  Include them in the introduction, the launch and the continual refinement of your offers. You need customers to collaborate with you throughout the offer lifecycle.

- *Get to know your customers.*
  Invest time in your customers, and allow your customers to invest time with you. This is needed to create the overall partnership. You need to understand your customers to know what they are looking for. There are six key attributes a customer will inherently look for during any interaction with the organisation.

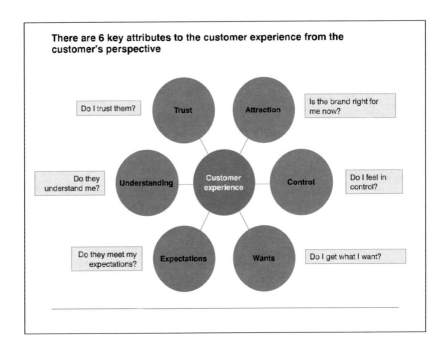

There are 6 key attributes to the customer experience from the customer's perspective

- Trust — Do I trust them?
- Attraction — Is the brand right for me now?
- Understanding — Do they understand me?
- Customer experience
- Control — Do I feel in control?
- Expectations — Do they meet my expectations?
- Wants — Do I get what I want?

- *Listen to your existing and potential customers, or their agents, on a proactive and reactive basis.*
  This is more than a one off event, it has to be continuous.

- *Show your customers you are listening.*
  This is not lip service. Customer feedback has to be delivered to the right person in the right department.

- *Act on the feedback you are given, treat it as a gift.*
  This involves, where relevant, incorporating customer suggestions into business processes and operations and keeping customers informed regarding their suggestions, even if the suggestion is not used.

All of the organisations detailed in the following case studies are collaborating with their customers, each in different ways.

A&R have assessed their route to market and have recognised that, as the company does not sell their product direct to the public, they are reliant on receiving favourable coverage from dealership and specialist press. Hence it is the dealers, with support from the specialist press, that are A&R's primary partners.

A&R have also recognised that if they relied solely on their close relationship with their primary partners they could have a biased view. As a result A&R use all their contacts to develop their products. They use their dealers and distributors to brainstorm new products; they use a small group of end customers, on proactive and reactive basis, for feedback on product ranges; and they use their suppliers and marketplace scans for insights into new technology.

A&R are willing to try new things and they accept that you can't always get it right first time. The beauty of their relationship with their extended partner network is that they should know that the product is right before it is hits the dealerships.

BAE SYSTEMS have turned their relationships with their customers to their advantage. They recognised that using customers in the later stages of product development was not enough as they needed to be involved in the process at the research stage. This recognition has led to a ground-breaking product, the silicon gyroscope, being developed that has been used in markets not considered by BAE SYSTEMS.

BAE also learnt some very important insights into the new market by engaging with their customers – for instance the customer requirements and the cost that the market would bear were known up front, plus, very importantly, the preferred route to market. This has led BAE to develop a new relationship with an intermediary to develop and launch the gyroscope.

Again BAE have used partner network to ensure that the product is right before it is developed and marketed.

Creature Labs are in a market that has traditionally seen short customer relationships. Creature Labs has taken the decision to engage with its customers in a number of different ways. They have, quite literally, allowed their customers to develop the product themselves. This has led not only to product innovation but longer term ongoing relationships. Customers who choose to engage with Creature Labs in this way are customer advocates. Creature Labs have let the customers take control of the relationship and in doing so have developed mutual levels of trust.

In addition, owing to Creature Labs' relationships with other businesses, in particular with education partners, they have been well placed to exploit market trends. These business relationships combined with the customer insights have led to truly innovative products.

In conclusion, whilst some organisations see direct contact with customers as the necessary evil of doing business, others embrace such contact and use it to develop and continually refine their offers. This is necessary as such engagement helps build customer loyalty and more importantly the partnership. At the end of the day, customers will invest time in the partnership if they can see that it creates value for them.

case 1

putting on the brakes

'It wasn't about selling what we could produce, it was about producing what the market wanted.'

When drivers slam on the brakes, they are more likely to pull safely out of a skid thanks to BAE SYSTEMS' innovative silicon gyroscope senses. 'It happens so fast that you can't detect it,' says Colin Fancourt, head of business for solid-state sensors. 'It spots your skid and gets you out of it before you realise there's a problem.'

Launched in 1997 for use in automotive braking systems, BAE SYSTEMS' gyroscope has won a significant share of the market. Applications in airbags and in-car navigation systems are being investigated, as well as uses in intelligent shells for the military and a 3D mouse for PCs.

These were nearly markets that BAE SYSTEMS missed out on altogether. By the mid-1990s it was becoming clear that classic gyroscope technology was not up to meeting the potentially massive demand for an inexpensive, rugged application in car braking systems to increase road safety. BAE SYSTEMS found its competitors were already thinking about making such gyroscopes out of quartz. So it decided to go for what appeared to be the far riskier option and set about creating the world's first silicon gyroscope.

Customers were included early on in the development of the silicon gyroscope. By asking them 'if we produced such a device, would you be interested?' BAE SYSTEMS found it was not alone in seeking to refine existing technology. Car companies such as Ford, General Motors and Rover wanted to improve braking systems using shock-resistant, inexpensive gyroscopes. But measuring the benefit of improved braking against consumer cost meant that existing technology could not be used. BAE SYSTEMS used its contacts to establish a price that the market would bear and that became its target. As it was a new product, buyers remained cynical, but their response was used to refine prototypes.

BAE SYSTEMS prides itself on its intimacy with customers. It devotes time and money to ensuring that it really delivers benefits. On the gyroscope project, BAE SYSTEMS for the first time took the backroom boys in research and brought them into the front room of marketing. 'We got them to work for the marketing manager,' says Colin Fancourt. 'An empathy quickly developed and soon they realised there was no value in developing a product the market didn't want. That it wasn't about selling what we could produce, it was about producing what the market wanted. So we took the researchers along to meet the buyers. It was a great communication exercise that reaped considerable benefits.'

Preferences among target customers in the automotive industry altered the BAE SYSTEMS strategy for producing the gyroscope. 'We learned that they were not interested in buying from an aerospace company,' says Colin Fancourt, 'they wanted to use a first tier automotive supplier.' BAE SYSTEMS switched direction and selected Philips Automotive to help in this area, ensuring the successful development and launch of the silicon gyroscope.

They are not merely customers, but rather disciples for the game – far more effective than sales staff.

Creatures is an unusual computer game. It creates artificial life, running several different neural networks, each one representing a creature's adaptive brain.

case 2

customers and disciples

As well as filing for a patent for its unique technology, the developer, Creature Labs in Cambridge, has encouraged a more informal flow of ideas in refining the game.

The game has a vocal fan club. Suggestions and comments flood in over the web. There are 175,000 visitors a month to the five active Creatures websites.

'Users are so adept at creating add-ons to the game,' says Chris McKee, the company's chief executive, 'that we thought it was worth encouraging. We have made it a mission to provide them with as many tools as we possibly can.' A virtual developers group was set up. Users become utility developers, taking a cut of the revenue, covered by a contract that gives the IPR rights to Creature Labs.

Short customer relationships are the norm in the games market, but the website has set up longer-term, ongoing relationships. 'Our customers know us better than our publisher.' They are not merely customers, but rather disciples for the game – far more effective than sales staff.

At each stage of its development, Creature Labs has been well positioned to pick up on wider market trends. Its early work in artificial intelligence found a use in schools through Acorn and the BBC, just as the national curriculum was taking off. Work on home education and computer games led to a contract with Olivetti, following a chance meeting.

Subsequently demand from Longman, a leading educational publisher, for a presence in European software set the company on a course as a distributor. These contracts and combinations have pulled together Creature Labs' capabilities in artificial intelligence and education to create a unique games technology.

case 3

wired in

Brainstorming is carried out not just internally, but with distributors and dealers, in order to identify new products.

From a site outside Cambridge employing 110 people, A&R produces hi-fi electronics under the brand name Arcam, occupying the middle ground between mass-produced and top-of-the-range equipment. It is a position that the managing director, John Dawson, jealously guards.

A series of good press reports in the late 1970s provided the foundation of A&R's initial success with dealers. Relations with the press and dealers are absolutely critical and matter more to A&R's prospects than relations with end users.

The Alpha 10 was designed in the mid-1990s to be a modular system to which different functions can be added. This new modular approach provides flexibility and offers a solution to year-by-year changes in technology. The idea was that customers would clearly be able to see an upgrade path.

The company website is used to promote products, but not to make sales. It is designed to warm up customers and drive them to the dealers. A&R actively trawls internet news groups and bulletin boards looking for trends and has an e-mail dialogue with 20–30 customers a day. Constant revisions are made to marketing strategy in the light of information received.

Brainstorming is carried out not just internally, but with distributors and dealers, in order to identify new products. New technology is gained in several ways – by keeping in close contact with leading component suppliers, by observing trends in related marketplaces (eg broadcasting, telecom, professional) and occasionally by reverse engineering Japanese equipment. Any relevant technology is either improved or licensed. The resulting innovation means that prices can be increased considerably.

More formal sessions with dealers are being considered to advise on product development. A&R spends £15,000 a year on market research, providing weekly and monthly reports, which John Dawson reads personally. 'A lot of the external data is dodgy, so it has to be used with great care', he says. 'I have become a world expert here. That's part of my job.'

A&R gives its staff a chance to try out ideas. In two cases engineers were employed to test out new approaches. They proved to be 'glorious failures', but there is no regret that the company took the risk.

The company has no patents. Dawson feels it is an open industry and he is aiming to set a standard for others. 'It's very difficult to protect circuit design. If someone did copy it, what could you do about it anyway?' He is far more committed to protecting the company's trademark.

A&R uses the brand name Arcam to avoid any confusion with AR (Acoustic Research), which was formerly based in Cambridge, Massachusetts, and with Cambridge Audio (part of the Richer Sounds group). It currently sells 35–40 per cent of its output overseas. It is aiming to lift that proportion to 50 per cent, although there are problems with currency and cheap Chinese competitors to be addressed. Difficulties are also experienced with constant demand for changes and upgrades. 'That is counter-productive to making things more reliable,' says Dawson 'but with the current ever-increasing pace of technological change it is absolutely necessary that we stay up with the world leaders in the field. Dealing with this efficiently is a major challenge for A&R, our dealers and end users.'

3.2 information systems: innovation – it's changed!

'Adaptive IT is able to constantly change to deliver innovation in the form of new processes in response to external demands, rather than the previous role of IT in enabling internal functions'

Andy Mulholland,
Chief Technology Officer,
Cap Gemini Ernst & Young

contributor profile

andy mulholland

**chief
technology
officer**

Andy Mulholland is Chief Technology Officer at Cap Gemini Ernst & Young. His role includes advising the company's management board on all aspects of technology promoted by market change.

Andy works with board level customers to help them understand the internal & external effects of technology on their business.

This covers both the internal and external effects on Cap Gemini Ernst & Young's customers and for their own business needs.

He has particular expertise in the alignment of technology business goals with emphasis on improving the communications and use of information in business processes. In a career spanning 32 years Andy has worked across all major industry sectors delivering business advantage for clients from new technologies.

He has been the founder or cofounder of four technology companies that have been the subject of either acquisition by leading multinational technology companies, or have gone public on NASDAQ.

To solve these problems Marks & Spencer pioneered Imagemaster, a 'colour by numbers' approach to colour communication. The principle behind this uses a special machine called a spectrophotometer to shine all the colours of the rainbow at a dyed fabric and measure how much of each colour is reflected. This gives the colour a numerical 'fingerprint' that can be faxed or e-mailed around the world.

All Marks & Spencer colour standards are now in data format – fabric suppliers now compare their dyed fabric to the standard colour at Marks & Spencer. A simple computer programme calculates if the colours match. No swatches need to be exchanged between Marks & Spencer and the fabric supplier. The approach was so successful that many other retailers copied this method of working to get the benefits of reduced lead times and reduced colour variation in store.

The only drawback of the colour by numbers approach is that it often operates in an invisible world where people have to deal with numerical descriptions of colours rather than having anything to see. Interpretation of the numerical data requires expert knowledge.

A system was therefore developed to bring the colour data to life by displaying it as true-colour on a calibrated screen. This enables a dyer in a remote location to e-mail a fingerprint of that colour to M&S where a designer can immediately view the actual colour on screen.

In addition to the quality control aspects of the system, software developments now mean that colours can be viewed on any textile or garment on screen, enabling designers to develop ranges of merchandise in true colour without the need for any actual dyeing to take place.

The real value for Marks & Spencer lies in improving its partnership with suppliers and cutting time to market. Whereas previously it used to take 18 months for new fashions to reach the shops, it is now expected that the process takes less than 12 months. Imagemaster is pivotal in delivering this. Clothing is the priority application for Marks & Spencer. It accounts for half of its £8 billion turnover, and 80 per cent of Marks & Spencer's clothing lines are in single colour fabrics suited to Imagemaster.

3.4 design and innovation

'today, most successful companies are using design as a way to help them differentiate their products, services and customer experiences'

**Jane Pritchard,
Innovation Services
Leader, IDEO**

contributor profile

jane pritchard

**innovation
services leader**

Jane joined IDEO London in the Autumn of 1999 to lead London's Innovation practice. IDEO is a world leader in the user-centred design of products, services, and environments. IDEO has won more design awards than any other firm of its kind and employs over 360 people in offices serving North America, Europe, and Asia.

During her time at IDEO Jane has carried out a variety of Innovation programmes with clients such as British Telecom, Lego, Nestle and Nokia. She is currently leading a series of ongoing Design & Innovation Workshops at London's Design Museum which are focused on transferring IDEO's innovation tools & techniques to a design education audience.

Jane has 7 years international experience in the field of Design Management. Before joining IDEO Jane spent 5 years working as a designer & project manager for a leading product development consultancy based in Switzerland. During this period, Jane was initially responsible for recruiting a design team and successfully implementing design as an 'in house' function. She worked on major projects for clients including Tetra Pak Japan, Bally and Ericsson.

Jane holds an undergraduate honours degree in Interior Design from Ryerson University in Toronto, Canada and an MA in Design Strategy & Innovation from Brunel University, UK.

3.4 design and innovation

Innovation is one of today's hottest business topics. Globalisation, maturing markets, deregulation, new technology, ecological constraints and more knowledgeable customers can put companies in unfavourable situations almost overnight. Continuous innovation is the key to sustaining competitive advantage in today's fast moving and global marketplace.

So, how does design fit into the innovation process? Design is everywhere – wherever you are, look around – everything that you see has been designed: a building, telephone, printer, light fitting, table or flower vase. The only thing that has not been put through some sort of design or innovation process is nature. Today, most successful companies are using design as a way to help them differentiate their products, services and customer experiences.

The role of design in the innovation process is as follows:

- Design research techniques can be used to help identify new product or market opportunities.

- The design process is used to both generate ideas and to implement solutions for useful, usable and delightful products, services and environments.

- Design techniques can be used to communicate innovation and new ideas through prototyping and visualisation.

Having good ideas is easy but actually getting them to market is a different matter. The following two case studies (Dyson and Linear Drives) provide examples of innovative companies that have used techniques common to designers and the design process, to help them achieve their business success.

Risk a little, gain a lot

Innovation by definition is about doing something new, and doing anything new implies a level of risk. However, in today's dynamic economy, not taking risk is the biggest risk of all. Companies must innovate – but they must carefully manage their risk in order to avoid costly mistakes.

When Dyson started working on his new product, he was told that there was not a market for it. Thus it became even more important for him to mitigate the risk. One way that Dyson did this was through prototyping and visualisation – throughout every stage of the design process. Dyson and his team prototyped early and often – they tested and evaluated their ideas – continually learning, iterating and collecting feedback in order to inform their next move. The idea for Dyson's DC03 product was actually generated by feedback on earlier product models.

As design teams, we need to test ideas early in order to reduce risk. Prototyping and visualisation techniques are extremely valuable as ways to minimise risk when developing new product or service ideas. Prototyping serves as a way to bring ideas to life, to keep them alive, make them tangible, and above all make them communicable. It can enable ideas to be taken rapidly to a point where they can be reviewed and evaluated quickly by a variety of audiences – including management teams and end users.

Prototyping must start early, and there needs to be many prototypes made – and thrown away – as part of the development process. Dyson built his first models using basic materials such as cardboard and polystyrene – materials that are not too precious and require little financial investment.

Visualisation techniques, including computer-based renderings, physical models, storyboard illustrations or video are used to portray life with future products or services before they even exist. Linear Drives Limited use video footage to help communicate their products to a broad audience at exhibitions and trade shows.

Prototyping and visualisation go hand in hand with innovation. They are ways of learning and improving the quality of the ideas – and thus reducing risk.

Interdisciplinary teamwork and collaboration

Design consultants must generate compelling and inspiring ideas – they must also solve problems. The business of both generating inspiring ideas and solving problems is not discipline specific but rather an experimental and collaborative process, and thus is it important to work in small interdisciplinary teams. At IDEO our interdisciplinary teams include experts with a variety of backgrounds: industrial design, interaction design, prototyping, human factors and engineering, as well as people with MBAs and business backgrounds.

Teamwork is central to the working practices at Dyson. It is through cross-functional teamwork that people with different skills and expertise can contribute their diverse perspectives, views and experiences to challenge the problem at hand. Innovation leaps across such different points of view. At Dyson, teams are flexible – they change in both size and mix of people as a response to each stage of the development process. Team members rotate from project to project in order to cross-fertilise ideas and share know-how. The design and innovation process is strengthened through the power of such collaboration and synergies.

Another core skill for innovation is the ability to manage external collaborations and to form working partnerships or strategic alliances with other companies whose skills are complementary to your own. Linear Drives Limited have demonstrated this by creating a network of distributors around the world – such collaboration is helping them to establish themselves in the market as a global player. It is worth mentioning that this approach also reduces risk – forming such a network is a useful way of testing or 'prototyping' the market without having to incur significant costs.

Connecting with customers and key stakeholders
Finding the right place to innovate can be challenging. As industries mature, it is no longer possible to differentiate a product or service on the basis of technology or even on price.

Some companies use market research to drive their next new product or service development. Dyson and his team did not rely on market research when deciding to pursue their new product idea. In the early days, when the idea was still embryonic, they were told that there was no market for their product. James Dyson and his team decided to pursue their idea anyway, believing that it would address an untapped market need.

A common problem with trying to use traditional quantitative market research, to create radically different products, is that the customers' ability to guide the development of new products and services is limited by their experience and their ability to imagine and describe possible innovations. In other words, customers have difficulty articulating unidentified, or latent needs.

To innovate from the perspective of the user or customer requires a deep understanding of the user's explicit and, more importantly, their latent needs. This understanding of customers latent needs can be gained through the use of specialist design research methods – methods that look at the behaviours of real users in real environments.

At IDEO, design teams carry out observations of real people in real life situations to find out what makes them tick and whether they have latent needs that are not being met through current products or services. The advantage in having the design teams collect these insights first hand is that they have a high knowledge of various technologies and are skilled at interpreting the insights and translating them into design concepts.

Now that Dyson's product is on the market, the company continues to improve their products by using feedback from a range of people – including end users as well as other important stakeholders such as repair experts and retailers.

This is also seen in the Linear Drives case study – they work closely with their distributors to understand and respond to feedback from international customers. Indeed, it was connecting with these customers that helped influence the company's design and raise it to international industry standard.

Conclusion
To innovate and to get a new idea to market requires risk management, teamwork and collaboration, lots and lots of prototyping, refinement and iteration. Innovation requires connecting with your customers and end users but also other key stakeholders – like repair experts and retailers.

Both Dyson and Linear Drives Limited demonstrate how two very different companies have proven that these activities are key to achieving innovation. It is not easy, but in today's business climate, companies have little choice.

At IDEO a user-centred design methodology is the basis of all work. It has five basic steps that are interpreted differently for each project according to its nature. They are:

- **Understand** the market, the client, the technology and the perceived constraints on the problem. As the project evolves, the constraints are often challenged, but it is important to understand them at the start.

- **Observe** real people in real life situations to find out what makes them tick: what confuses them, what they like, what they hate, if they have latent needs not addressed by current products and services.

- **Visualise** new-to the-world concepts and the people who will use them. Some people see this step as predicting the future, and it is probably the most brain-storming intensive phase of the process. Visualisation can take the form of a computer-based rendering or visualisation, though IDEO also builds thousands of physical models and prototypes. For new product categories the customer experience is sometimes visualised using composite characters and storyboard-illustrated scenarios. In some cases a video is even made which portrays life with future product or service before it even exists.

- **Evaluate** and refine in a series of quick iterations. We try not to get too attached to the first few prototypes, because we know they'll change. No idea is so good that it cannot be improved upon. We get input from our internal team, from knowledgeable people not directly involved with the project, and from people who make up the target market. We look for what works and what doesn't and we incrementally improve the product in the next round.

- **Implement** the new product for commercialisation. This phase is often the longest and technically most challenging in the development process, but the ability to implement successfully lends credibility to the creative work that goes before.

This simple process works for everything from creating children's toys to launching new financial services. It places users, or customers, in the centre to ensure that the technologies employed serve their purpose.

When a group of engineers started working on a new design for vacuum cleaners in a coach house at the back of James Dyson's home, they were told the market did not exist and the selling price would preclude success.

The company now turns over £223 million and boasts the top nine models in the UK market[1].

Dyson takes risks. The DC01 design bent too many rules for others to risk providing capital or support. The DC03 targeted markets that did not previously exist; it was 'a complete leap of faith.'

Flexible, multi-disciplinary design teams are set up for each model development. They grow as the project progresses through its development process and the same team stays with the product after it is launched to support continuous design. Teams working on different projects rotate for cross-fertilisation of ideas. 'Teams give

case 1

a 'non-existent' market worth £223m

[1] GfK Lektrak May 2001 (by value).

Dyson its spirit,' says James Dyson. 'The teams need to be at the heart of the company. They are engaged in the core task of making a difference, through creating and selling new products.'

Risk is managed through a structured and phased development process based on prototyping, testing and evaluating at all stages. First models are built using cardboard and polystyrene, and then key components are tested. Each stage has numerous iterations and formal reviews. The setting of targets for materials controls costs. Tooling costs are amortised over expected sales forecasts. A sizeable test facility is used to help development and testing against competitors' products – as well as against recognised standards.

The idea for the DC03 came from feedback from the earlier models. James Dyson and the engineering design team then identified what they believed to be a niche that was not being addressed. Marketing has an input to the initial design specification, which 'captures the core values of the product'. The helpline and in-house repair service provide a stream of improvement data and retailers are also asked for input.

case 2

linear thinking

Essex-based Linear Drives Limited currently holds 5–10 per cent of the rapidly growing world market in industrial linear motors but predicts taking 20–30 per cent within three years.

Managing director Bill Luckin never tires of underlining innovation to his staff as the principal means of ensuring commercial survival: 'The only defence from competition,' he says, 'is to be out there first and to be the best.'

The company is clear in its ambition to establish a worldwide brand in its field and to build a suite of patents round its core technology. 'People are going to copy our product,' says managing director Bill Luckin. 'But we've got a head start now and we are determined to capitalise on it.'

As a small player that must sell its products on a global market, Linear Drives Limited has sought ways of extending its reach while reducing risk. It has therefore developed a set of strategic alliances and appointed distributors to sell its products into new markets. A mixed marketing strategy includes advertising, the use of a company website and attendance at exhibitions, where Linear Drives Limited employs video footage to help explain its products. It works at interrogating distributors in order to respond to feedback from international customers. It was feedback from customers that determined the company's shift from its traditional Phase 2 technology design to the international industry standard, Phase 3 technology.

Linear Drive Limited's major markets are Germany and the US. But it also has distributors in Sweden, Holland, Denmark, Italy, France, Belgium, Luxembourg, Singapore and South Korea. Linear Drives Limited is careful to get good distributors in place. It is prepared to forfeit sales rather than spread itself too thinly with distributors who are not appropriate. 'We've got people banging down the door wanting to represent us,' says Luckin. 'But the task is to pick out the right ones.'

3.5 working with rules and standards: rewriting the rules on regulation

'companies who are great at innovation recognise that regulation can actually be a very powerful driver for innovation itself'

**Nigel Crouch,
Senior Industrialist,
Innovation Group, DTI**

contributor profile

nigel crouch

senior industrialist

Nigel Crouch is a Senior Industrialist, who has run businesses for the past 19 years and been committed to innovation throughout his career. He has held a number of senior marketing positions with Cadbury Schweppes, both in the UK and internationally, and was on the Board at Reckitt & Colman as New Business Director of their Household Products Division. He moved into general management with Ciba and was then Managing Director of the Evo-Stik Adhesives & Sealants business of Evode Group for almost nine years. He currently runs his own investment business and spends part of his time working with the Innovation Group of the Department of Trade & Industry.

At the DTI, Nigel is heavily involved in several major on-going leadership and innovation programmes, working closely with a comprehensive network of external partner organisations. In conjunction with the Design Council, he has played a pivotal role in both the fieldwork research and roll-out of the powerful 'Living Innovation' findings, which are being very positively received by businesses across the UK. He has also been very involved from the DTI side in the development of the '100 Best Companies to Work For' list, published annually by *The Sunday Times*, and based upon an independent evaluation of the real perceptions people have of the organisations they work for and, allied to this, helped lead 'Partnerships with People'. This is part of the Fit for the Future campaign and one of the most in-depth investigations of recent years into how a number of highly successful organisations have managed to bring the best out of their people to significantly enhance their bottom-line performance.

Copies of 'Living Innovation', Partnerships with People' and the '100 Best Companies to Work For' can be obtained free from DTI, Admail 528, London SW1W 8YT or on orderline telephone 0870 150 2500 and fax 0870 150 2333. Additional information is available at www.dti.gov.uk/pwp; www.livinginnovation.org; www.lidiagnostic.com; www.sunday-times.co.uk/100bestcompanies

working with rules and standards: rewriting the rules on regulation

When the words 'regulations' and 'standards' are mentioned, many businesses shudder. They then launch into an emotional outpouring on the frustrations of having to deal with a plethora of red tape and legislative complexity on top of all the other problems that they face and urge the Government 'to do something about it'.

This reaction is very understandable and there is undoubtedly a real onus on Government to get really close to businesses on these difficult issues that have, of course, been further exacerbated by the rapidly escalating European dimension. For the UK to increase its productivity and achieve its rightful place on the world stage, Government needs to make its vital contribution to UK innovation and take an innovative approach itself on simplifying and cutting back on the regulatory structure and framework wherever possible.

A key element in this innovative approach is Government connecting with its business customers in the same way that highly innovative businesses connect incredibly closely with their customers. It is getting right inside the head of businesses, through close and regular contact at different levels, to really understand the problems they are facing and then come up with creative solutions that provide a 'win-win' result for both sides.

But it is important to note that the onus is not just on Government. There is, equally, a very real responsibility on the part of businesses to adopt a much more positive and pro-active stance on regulation than is, unfortunately, often the case. Companies who are great at innovation recognise that regulation can actually be a very powerful driver for innovation itself.

The success of Microsense Systems Ltd with their new and very innovative LED traffic light system can be directly attributed to the fact that they involved the Highway Agency right from the outset and at every key stage in the development process. Crucially, this enabled them both to meet the very stringent specification standards that were being applied and to move very quickly to market when critical technical breakthroughs were achieved. 'We are the only country that has these specifications and approval processes but it does mean the quality is better than anywhere else in the world … and it gives us a flag that we can use around the world.'

Gecko also had to meet extremely demanding specifications that were imposed by the RNLI during the development of their new marine safety helmet and this, again, was a key factor in their ultimate success. These stretching standards did give them a lot of headaches: 'It would have been easy to give up. On several occasions, I felt like saying I would get more money packing shelves!' But they showed that they had the necessary 'innovator characteristics' of determination, doggedness, courage and absolute commitment to win through against larger competitive rivals. Moreover, they then very positively turned regulation to their own advantage. Before the development of their marine safety helmet, there was no approval standard for helmets of this type but now, having worked closely with BSI, the Gecko product has effectively set the standard and become the international benchmark, thereby giving Gecko substantive competitive advantage. Rather than complaining about the restrictions legislation and regulation place upon them, let's see companies getting positively involved with Government at an early stage and working together to establish mutually beneficial principles, guidelines and standards.

This principle of a regulatory and legislative framework providing innovation opportunities was also manifest at the Bank of Scotland, when they developed their pioneering Shared Appreciation Mortgage. The release of cash for equity in the homes of older people is an area fraught with ethical and legal difficulties but the Bank of Scotland were able to successfully overcome the hurdles through the powerful co-working of the legal and commercial teams with strong top management support.

One of the most exciting opportunities for innovation in the regulatory area is related to environmental issues and sustainable development. If businesses can start to embrace a much wider community role and become much more pro-active in terms of environmental protection, they can generate very significant and beneficial differentials in their consumer propositions with consequent positive impact on their bottom-lines. Over and above this, however, they will also be able, in many cases, to pre-empt legislation and regulation through a strong and mutually beneficial voluntary code – business and Government working hand in hand for the common good with minimum statutory controls.

Innovation is all about fleetness of foot and a genuinely totally inclusive approach in terms of all the relevant parties both within and outside the organisation. It is also all about fundamentally re-writing the conventional rules on regulation and the potential rewards for both sides are extremely significant, if both business and Government are willing to bite the bullet and go for it.

①

Sacree has turned regulations to his benefit.
For many years Jeff Sacree of Gecko worked happily in the sleepy town of Bude in North Cornwall producing fibreglass surfboards. But during the cold winter months the surfers stayed at home and business was very slow.

case 1

becoming the stardard

He conceived the idea of creating a surfer's helmet that would not only protect its wearer, but also reduce the loss of heat through the head. This would allow surfers to stay in the sea for longer – and hopefully extend the summer surfing season.

Sacree had originally intended the Gecko helmet for use only by surfers. He was inspired to explore other applications through a chance conversation. A member of the local lifeboat crew explained they used motorcycle helmets that were bulky, cumbersome and reacted badly to salt water.

The RNLI confirmed it was looking for an alternative to the motorcycle helmets being used at the time, but explained that three bigger companies with more than 100 staff each were working on creating a helmet for use on the lifeboats. Sacree was undeterred and set about creating the first of many prototypes.

The RNLI proved to be very demanding in its requirements. The rival companies were asked to make many modifications over the following months. Being a one-man band, Jeff could focus exclusively on the Gecko helmet – unlike the bigger companies which all backed away from the project. Total commitment was essential, because it was two years before the RNLI finally placed a contract to manufacture the helmet. 'It was very hard work with very little return and a very shaky foundation,' says Sacree, 'but the RNLI are demanding and this is a very good benchmark.'

Sacree has turned regulations to his benefit. Before the development of the Gecko Marine Safety Helmet there was no approval standard for helmets of this type. The Gecko was used by the BSI when formulating standards and has become the benchmark for products in this field. Sacree is seeking to reinforce this competitive advantage by gaining ISO 9000 accreditation.

Gecko has the flexibility to create helmets for a wide variety of applications. Jeff is responsible for both client liaison and design, so customer feedback can be immediately reflected in product development. The company now employees four people to make 80 helmets a week for niche markets ranging from lifeboat crews and surfers to river police, customs officers and helicopter winchmen.

Because the helmet is close fitting and very light it can be worn for long periods. Since it is handmade it can be adapted for specialist uses, including the fitment of altimeters and video cameras to meet the needs of skydivers. In some client organisations, users have to keep a record of usage. This has proven very useful in making further improvements.

Organisations such as the Coast Guard and many private individuals in America have made contact with Gecko through its website, but the company is not considering expansion into the US market because of fears of exaggerated litigation claims.

The surfer's helmet is sold through specialist retailers. All other helmets are sold direct, with payment by credit cards and distribution through a carrier. This reduces investment in stock and limits exposure to changes in demand patterns.

Gecko has not invested in equipment and machinery to automate the production process. This has not only minimised financial risks but also enabled the company to retain the flexibility to produce helmets for tiny niche markets, thereby defending its market position from bigger competitors. 'It's a case now of looking at niche markets that are not big enough for a large manufacturer,' says Sacree.

Business Links have provided 'exceptional' support and advice on retaining ownership of the company, patenting and many other issues. The Gecko trademark has been registered in 15 European nations.

Sacree originally funded development of the Gecko Marine Safety Helmet through a bank overdraft. The overdraft was extended to pay staff wages when the RNLI contract was in place. Jeff is exasperated by the difficulties encountered in the very early stages of business development. His local bank knew his business well but was not permitted to take lending decisions. Investment offers only came from private investors when success was guaranteed, but these were turned down on the advice of the local Business Link advisor. Slow payment from customers remains a problem.

Microsense Systems Ltd was started 20 years ago from a Portakabin. Now it has embraced new technologies, gained UK and European approval and cornered an unique market with its new safer, low-maintenance, environmentally friendly traffic light system.

The company designs, produces and maintains electronic equipment for road traffic control. When light emitting diode (LED) technology was developed, the company immediately spotted its huge potential for use as a replacement for the conventional halogen bulb in traffic light systems. LEDs last longer, use less energy and have improved safety features.

The system was developed with the full involvement of Microsense's customers. The Highways Agency helped in the project from initial input of ideas and designs to the final testing and the ultimate approval of the product.

By getting the product to market quickly and successfully meeting the Highways Agency's stringent specification standards, Microsense has cornered a niche market for new systems and is beginning to break into the market for replacing existing units. The potential for expansion into other markets, in the UK and abroad, also looks promising.

During the development the Highways Agency was fundamentally involved in product development, testing and approval. The initial idea for the product came from both customers and in-house engineers and was strongly promoted by the top management. The company also worked with its customers to think longer term. While the cost of replacement is high, energy savings, reduced costs of failure and environmental benefits produce a payback within nine years. Microsense is working to reduce this period.

Its biggest market, however, still lies with their existing halogen units. When customers see the new technology it can provoke a lot of very interesting suggestions and ideas, some of which are already being pursued. Having a good understanding of the technology also allows Microsense to explore other markets for the traffic signal.

The specifications drawn up by the Highways Agency and standards for traffic lights in the UK are exceptionally high and were a tangible limiting factor. The colour and intensity of the lights has to meet very strict criteria. Whilst red and orange LEDs were available, Microsense had to wait for manufacturers to produce a suitable green LED.

Eventually, these restrictions worked to the company's advantage; because they were closely following developments, they were able to move quickly when the green light was produced. Having gained UK approval, they can now use it as a stamp of quality around the world. Meeting the strict UK specifications works in the company's favour when addressing markets with less demanding requirements.

Appendix

This self-assessment tool is based on Living Innovation and is a brand new development from the DTI's Innovation Group and the Design Council, which is currently being piloted and refined. This can be viewed on www.lidiagnostic.com

living innovation: self assessment

Successful innovation is the key to improved business performance. The Design Council's Millennium Products initiative recognised the very best in British creativity, innovation and design. As part of the initiative the DTI, working with the Design Council, carried out an in-depth investigation of the business processes adopted by more than 50 of these Millennium Products award winners. Living Innovation reports the findings.

The approaches and methods used by successful innovators in Living Innovation are the basis for this innovation self-assessment tool. It incorporates evidence from large and small firms, established businesses and start-ups from all sectors in today's knowledge economy.

This self-assessment tool is designed for use at any level, in any organisation. By completing it, you will be able to compare your organisation's culture, management, processes and performance with those of the leading firms investigated in Living Innovation.

Read carefully each of the sets of statements given, then select and mark the one that most closely reflects your own view of your firm's current normal practice. Remember – be honest, and try not to 'gild the lily'!

1. ➡   A❑ B❑ C❑ D❑

A   We ensure that people get basic skills training when it is required.

B   People are consistently trained and developed for bigger roles in the organisation.

C   People thrive on the opportunity for personal development that they created for themselves.

D   Training and development of our people pervades every part of our organisation.

2. ➡   A❑ B❑ C❑ D❑

A   Our potential for innovation is inevitably limited by the very necessary financial constraints applying.

B   One of our strengths is that finance is always available to exploit our good ideas.

C   Finance is made available for the majority of good ideas to be taken forward.

D   We see the availability of finance as a very necessary filter of good ideas.

3. ➡   A❑ B❑ C❑ D❑

A   We establish the needs of our customers mainly through regular face-to-face customer reviews.

B   We establish the needs of our customers by regularly carrying out postal surveys.

C   Our customers are involved very closely with our business so that we all fully understand their needs.

D   We see quality market research as the best determinant of what our customers want.

4. ➡   A❑ B❑ C❑ D❑

A   Managers lead by example and will always do what they say they will do.

B   It is understood that managers very often have to play their cards close to their chest.

C   Managers are committed and generally seen to be open whenever they can.

D   Managers know what they want but people sometimes question their motives.

5. ➡   A❑ B❑ C❑ D❑

A   People do sometimes work in teams here when the need arises.

B   We mainly operate with multi-functional teams made up of people within the organisation.

C   With the odd exception our teams here generally relate to specific functions.

D   Our teams are usually made up of a broad mix of people from both within and outside the organisation.

6. ➡   A❑ B❑ C❑ D❑

A   People are given considerable freedom within their own areas of responsibility to work as they want.

B   As long as they are contributing to the clearly defined objectives of the organisation, people are allowed to work in the away they see most appropriate.

C   Managers have to take the decisions but individual initiative is positively encouraged.

D   We operate here to tried and tested job methods that have worked very well for us.

7. ➡   A❑ B❑ C❑ D❑

A   We are focused on an incremental approach to innovation within our markets.

B   Our success is soundly based upon reacting well to market changes as they occur.

C   We actively look for opportunities for both incremental and quantum leap innovation in our business.

D   We have been able to transform a market we operate in and maintain it through innovation.

8. ➡   A❑ B❑ C❑ D❑

A   There are particular people in the organisation whose role it is to be informed about trends and deliverables in our industry.

B   Most of our people use tried and tested sources of information about what is happening in our industry and market.

C   Everyone is encouraged to go out and find external sources of inspiration and information to keep in touch with what is happening nationally and internationally.

D   We are able to track industry or market developments by reading relevant press.

9. ➡   A❑ B❑ C❑ D❑

A   We do not believe that we can protect our intellectual property on most of what we do.

B   Our intellectual property is at the core of our competitive advantage and we protect it accordingly in the most appropriate way.

C   We are generally confident in our competitive advantage and feel there is usually no need for legislative protection.

D   We always look to protect our intellectual property on really major breakthrough developments.

10. ➡   A❑ B❑ C❑ D❑

A   Legislation and regulation provide necessary constraints that we have to work within.

B   New legislative requirements very often provide us with competitive advantage.

C   We regularly exploit compliance with regulatory standards for business benefit.

D   We always seek to influence relevant new legislation and regulatory standards and benefit accordingly.

11. ➡   A❏ B❏ C❏ D❏

A   We have a statement of the vision and it is available for everyone to see if they wish to.

B   Everyone here is really excited and driven by the vision of the organisation.

C   Most people know our vision and what they have to do to help achieve it.

D   Managers have developed the vision and sought to disseminate it widely.

12. ➡   A❏ B❏ C❏ D❏

A   Top managers are very 'hands-on' but are willing to devolve some responsibility.

B   Top managers are 'hands ready' knowing exactly where and when not to get involved.

C   All decisions are taken at the top and what people do is very clearly prescribed.

D   The style here is very 'hands-off' with people left very much to do their own thing.

13. ➡   A❏ B❏ C❏ D❏

A   We are always seeking to improve our relationships with our suppliers and customers.

B   We rigorously and regularly monitor and review our relationships with customers and suppliers.

C   Reviewing our relationships with customers and suppliers is an important part of our strategic planning process.

D   We deal with our suppliers as necessary when we need to address the needs of a particular customer.

14. ➡   A❏ B❏ C❏ D❏

A   We go about our business here in a quiet and purposeful way.

B   We have a good working environment and people are very friendly.

C   There is a tremendous buzz working here and people have fun achieving a lot.

D   Everyone here has a job to do and they are expected to get on with it.

15. ➡   A❏ B❏ C❏ D❏

A   We base the development of our products/services on a recognised design process.

B   We use design processes at the centre of our innovation strategy.

C   When we need to use design, we generally bring designers in from outside our business.

D   We do not see design as part of the innovation process.

16. ➡   A❑ B❑ C❑ D❑

A   We all help each other here and always get and offer whatever support is needed.

B   People here depend on our managers to solve their problems.

C   It is very important here not to impose on others and to be totally self-sufficient.

D   Managers are accessible and always try to help us when we ask them.

17. ➡   A❑ B❑ C❑ D❑

A   Opportunities we identify are driven by our deep understanding of our customers and markets.

B   Tracking market share is a particularly critical determinant of market dynamics.

C   We analyse the key drivers in our markets and then ensure that we respond in the appropriate way.

D   Comprehensive sales analyses combined with market surveys when appropriate provide the focus for new areas of opportunity.

18. ➡   A❑ B❑ C❑ D❑

A   There are no problems with the commitment or morale of people here.

B   People are passionate about what they do and really want to create something special.

C   We feel that people are as satisfied here as they would be anywhere else.

D   People are well-motivated and are generally very enthusiastic about their jobs.

19. ➡   A❑ B❑ C❑ D❑

A   Team leaders have a certain amount of freedom within well-defined boundaries.

B   Our teams provide an important input to day-to-day decisions by managers.

C   Teams here have a lot of discretionary power with strong support from the top.

D   Management decisions are not generally based on input from teams.

20. ➡   A❑ B❑ C❑ D❑

A   Most of our customer contact points tend to be with buyers or specialist functional people.

B   We are very good at maintaining high-level relationships with our customers.

C   With many of our customers there are a number of different customer contact points spanning the whole of both their and our organisations.

D   Where we have large customers we entrust client relations to an account manager.

21. ➡   A❏ B❏ C❏ D❏

A People are asked for their inputs but managers usually make up their own minds.

B Managers are very serious about our briefing and consultation process.

C Everyone is actively encouraged to come up with good new ideas and to help see them through.

D People are listened to in this organisation and their ideas are acted upon.

22. ➡   A❏ B❏ C❏ D❏

A Managers rigorously evaluate all ideas to minimise our exposure to risk.

B We always take a low risk approach given the difficulties inherent in evaluating risk.

C Managers here are always ready to take a certain degree of calculated risk.

D The effective elimination of all risk has been a critical factor in our success.

23. ➡   A❏ B❏ C❏ D❏

A When our customers have identified problems with our product/service, we involve a trained designer in looking at how we could improve.

B Both internal and external designers are involved in helping our business to think about the implications of future trends and issues for our customers.

C We usually only modify our products and services when a new technical development becomes available.

D The development of our products/services is driven mainly by our sales team who we feel know our customers best.

24. ➡   A❏ B❏ C❏ D❏

A We have good systems here for rewarding good performance.

B People are rewarded by being paid fairly for delivering what is expected of them.

C What is great here is that good work is always recognised and really appreciated.

D We have found that personal recognition runs the risk of being divisive.

25. ➡   A❏ B❏ C❏ D❏

A When problems occur, senior managers immediately act to provide appropriate support and guidance.

B When major problems arise, senior managers will then take strong control.

C Senior managers offer support when innovative projects are challenged but tend not to intervene.

D Senior managers like to get heavily involved with innovation projects and drive them.

26. ➡   A❏ B❏ C❏ D❏

A Our strength has been built on consistently delivering what we do best.

B People can take new ideas forward when the day-to-day pressures permit.

C There are significant parts of the organisation that are highly innovative.

D We are innovative throughout the organisation and this is crucial to our success.

27. ➡   A❏ B❏ C❏ D❏

A We regularly break new ground in our markets in order to maintain our competitive position.

B It is very hard for us to differentiate ourselves in what are essentially commodity markets.

C Our competitive advantage is firmly based on what we have always done well.

D We have tried a variety of ways to improve our competitive advantage with some success.

28. ➡   A❏ B❏ C❏ D❏

A We have very high standards and it is widely appreciated that mistakes cannot be tolerated.

B Our people prefer to work well within their capability rather than make a mistake.

C People work hard and managers are generally tolerant of the occasional genuine mistake.

D People stretch themselves and when we do make mistakes we learn from them.

29. ➡   A❏ B❏ C❏ D❏

A New ideas come from management and are passed down for development.

B All levels of the organisation are proactive in generating, evaluating and developing ideas for new products/services/processes.

C We are all encouraged to think and act creatively.

D   Creativity is generally considered to be the remit of a particular group of people in the company.

30.   ➡   A❑ B❑ C❑ D❑

A   We provide company information online and are now actively developing our e-business strategy.

B   We have an e-business strategy in place and make extensive use of the internet.

C   Our knowledge and utilisation of e-business has enabled us to fully exploit our potential.

D   Our customers do not currently use the web and we use other means to service them very effec-
         tively.

Are you content with all your answers? If so, the next step is to analyse your responses. Using the score
card on the next page, please enter your individual scores for the 30 questions in the results grid opposite.
Here your scores are arranged into the nine significant factors found in firms with successful innovation
cultures and processes. You will see that they are characterised by the key aspects of innovative behaviour
identified by the findings of Living Innovation and mirrored in this book: Inspire, Create and Connect.

living innovation

scorecard

Go back through your responses. Identify the score for each question and enter it on the grid on the next page.

1.		9.		17.		25.	
A	3	A	3	A	7	A	7
B	7	B	7	B	4	B	3
C	4	C	4	C	6	C	6
D	6	D	6	D	3	D	4

2.		10.		18.		26.	
A	3	A	3	A	4	A	4
B	7	B	6	B	7	B	3
C	6	C	4	C	3	C	6
D	4	D	7	D	6	D	7

3.		11.		19.		27.	
A	4	A	3	A	6	A	7
B	3	B	7	B	4	B	3
C	7	C	6	C	7	C	4
D	6	D	4	D	3	D	6

4.		12.		20.		28.	
A	7	A	4	A	3	A	3
B	3	B	7	B	6	B	4
C	6	C	3	C	7	C	6
D	4	D	6	D	4	D	7

5.		13.		21.		29.	
A	3	A	4	A	3	A	3
B	6	B	7	B	4	B	7
C	4	C	6	C	7	C	6
D	7	D	3	D	6	D	4

6.		14.		22.		30.	
A	6	A	4	A	6	A	4
B	7	B	6	B	3	B	6
C	4	C	7	C	7	C	7
D	3	D	3	D	4	D	3

7.		15.		23.			
A	4	A	6	A	4		
B	3	B	7	B	7		
C	6	C	4	C	3		
D	7	D	3	D	6		

8.		16.		24.			
A	4	A	7	A	6		
B	6	B	3	B	4		
C	7	C	4	C	7		
D	3	D	6	D	3		

living innovation

results

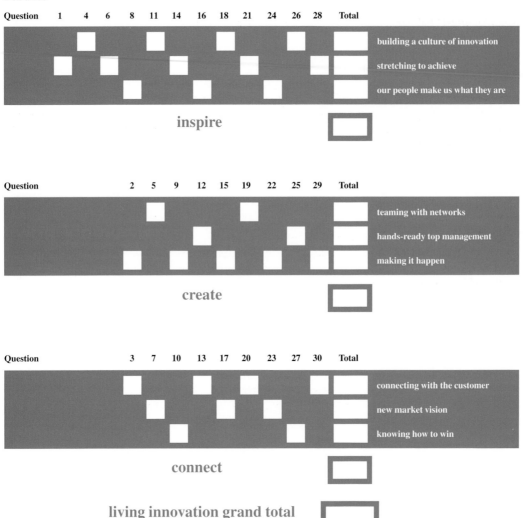

Question	1	4	6	8	11	14	16	18	21	24	26	28	Total	
		☐			☐			☐			☐		☐	building a culture of innovation
	☐		☐			☐			☐			☐	☐	stretching to achieve
				☐			☐			☐			☐	our people make us what they are

inspire ☐

Question			2	5	9	12	15	19	22	25	29	Total	
				☐			☐					☐	teaming with networks
					☐				☐			☐	hands-ready top management
			☐		☐	☐			☐		☐	☐	making it happen

create ☐

Question			3	7	10	13	17	20	23	27	30	Total	
			☐			☐			☐		☐	☐	connecting with the customer
				☐			☐		☐			☐	new market vision
				☐						☐		☐	knowing how to win

connect ☐

living innovation grand total ☐

Add up the scores in each row, giving you scores for the nine significant factors that emerged from the Living Innovation study. Then add each set of three as shown, to give you total scores for Living Innovation's key themes – Inspire, Create and Connect.

Finally, add the three together to give your Living Innovation grand total. Now turn to the conclusion to interpret your performance.